INNOVATION AND APPLICATION OF GREEN BUILDING
MATERIALS
— PUBLIC BUILDINGS

绿色建筑材料的创新与应用 · 公共建筑

VOL . I

深圳市艺力文化发展有限公司 编

华南理工大学出版社
SOUTH CHINA UNIVERSITY OF TECHNOLOGY PRESS
· 广州 ·

图书在版编目（CIP）数据

绿色建筑材料的创新与应用 = Innovation and application of green
building materials：汉英对照 / 深圳市艺力文化发展有限公司编 .-- 广州：
华南理工大学出版社，2014.4

ISBN 978-7-5623-4187-1

Ⅰ．①绿… Ⅱ．①深… Ⅲ．①建筑材料 - 无污染技术 - 汉、英 Ⅳ．① TU5

中国版本图书馆 CIP 数据核字（2014）第 044423 号

绿色建筑材料的创新与应用
Innovation and Application of Green Building Materials
深圳市艺力文化发展有限公司 编

出 版 人：韩中伟
出版发行：华南理工大学出版社
（广州五山华南理工大学 17 号楼，邮编 510640）
http://www.scutpress.com.cn　E-mail: scutc13@scut.edu.cn
营销部电话：020-87113487 87111048（传真）
策划编辑：赖淑华
责任编辑：史　册　赖淑华
印 刷 者：深圳市汇亿丰印刷包装有限公司
开　　本：965mm×1440mm　1/8　**印张：**60
成品尺寸：245mm × 330mm
版　　次：2014 年 4 月第 1 版　2014 年 4 月第 1 次印刷
定　　价：598.00 元（上、下册）

Preface

如今，很多建筑师是先创建建筑外观，再询问工程师如何施工以及使用何种材料。路易斯·康提出了疑问：砖块本身想要变成什么？我一向欣赏康对待材料的方式。我有一个近乎痴迷的习惯，当在小笔记本上画出建筑的第一条概念草图线时，它就已经包含了要使用的材料的构想。

若从材料选择的角度浏览本书的项目，那么很显然，在此方面，一千年来都未曾发生革命性的改变。不过，很多材料的使用方式，却有所变化。

新的技术解决方案也运用于玻璃，并且有色玻璃的使用也明显增加。借助计算机辅助数控切割，金属片可以变成任何形式的装饰，运用于立面的可视表面，且无需额外费用。与混凝土一样，钢材仍然是最重要的结构材料。

我的同胞，建筑师埃罗·沙里宁，最早于 1963 将耐候钢运用于建筑中，即美国伊利诺伊州约翰迪尔全球总部。之后，这一色调温暖、呈锈色的钢材几乎被遗忘，好在近年来，许多有趣的建筑中再度出现它的身影。

在未来，我们还可以看到利用纳米技术创造的新材料。2014 年索契冬季奥运会速度滑冰中心就是一例，它的纳米天花板，可以防止观众席的热量对场馆的冰质造成损害。

现在被广泛用于帆船和汽车行业的碳纤维，未来或许会在建筑中大放异彩。碳纤维具有强度比钢大、重量却轻得多的优点。

玻璃纤维增强混凝土，借助参数数字化建模和相关成型技术，可用于实现想要的形状。

若需更小的尺寸或更大的跨度，则可使用结构纤维混凝土来实现。

我有幸多次荣获"年度最佳混凝土结构奖"，其中四次在我的祖国领奖，一次在德国。首次获此殊荣是在 1983 年因拉赫蒂城市剧院，它是我的首座大型公共建筑项目。那时，混凝土作为建筑材料所提供的种种可能性，使我深深着迷，这座剧院也完全以清水混凝土建造而成。2003 年，我因设计哥特式红砖教堂的一座音乐厅而荣获德国混凝土建筑奖，该音乐厅再度使用清水混凝土饰面。我对这一材料的执著，源于我相信它足以媲美天然材料，建筑师可以如雕塑家一样筑造它，或如安东尼·高迪塑造天然石材一样打造它。

在追求可持续发展之路上，没有何种材料是被完全禁止的，尽管不同材料已按优先顺序进行排列。

木材得到越来越多的使用，这也许可以视作一个大趋势，即环保考量激发建筑师寻找新的表达形式。在一些国家，建造高层木制建筑的能力得到提升，这得益于相关研究成果以及消防安全法规的修正。不幸的是，在我撰写本序时，一场毁灭性的火灾将挪威一座历史悠久的木制村落烧毁殆尽，恐怕在短期内将给木制建筑带来负面的影响。尽管如此，新建筑物的建造，仍需以现代城市规划、材料开发以及安全法规的更新为主导。

显而易见，在未来，选择材料将越来越多地考虑到，它能在多大程度上支持可持续发展的目标。其中一个极好的例子，是覆盖整个立面的垂直绿化。

佩卡·萨米宁，教授，芬兰建筑师协会建筑师

PES 建筑设计事务所

There are a great many architects today who first create the form and then go and ask an engineer how to build it and what materials could be used. Louis Kahn posed the question: "What does the brick want to be?" I have always admired Kahn's approach to materials. It is almost an obsession with me that the first concept line for a building that I draw in my little Moleskin notebook already includes an idea of the material to be used.

When viewing the projects in this book from the perspective of material choice, it is obvious that nothing revolutionary has taken place in this respect in this millennium. However, the ways in which many materials are used have changed.

New technical solutions and uses have been developed for glass, and the use of coloured glass has clearly increased. With computer-aided CNC cutting, metal sheets can be turned into practically any kind of ornamentation in the visible exterior of the facade, without significant additional cost. Together with concrete, steel remains the most important structural material.

My compatriot, the architect Eero Saarinen, was the first to use Cor-Ten steel in architecture in the John Deere World Headquarters in Illinois, USA, in 1963. After that, this warm-hued, pre-rusted steel was largely forgotten, but in recent years it has appeared again in many interesting buildings.

In the future, we may also see new materials created with the use of nano technology. The Speed Skating Centre built for the Sochi 2014 Winter Olympics, for example, has a nano ceiling, which prevents the heat from the audience from damaging the ice.

Carbon fibre, a material now commonly used in sailboats and in the automotive industry, may well be used in architecture in some form in the future. Carbon fibre has the advantage of being tougher than steel while being considerably lighter.

Glass fibre Reinforced Concrete, GRC, can be used to realise demanding shapes with the aid of parametric digital modelling and the related moulding technology.

Smaller dimensions or greater spans can be achieved with the use of structural fibre concrete.

I have had the honour of receiving the "Concrete Structure of the Year" award four times in my home country and once in Germany. I first received this award for my first major public building, the Lahti City Theatre, in 1983. At that time, I was fascinated with the possibilities offered by concrete as an architectural material, and this theatre building is built entirely of fairface concrete. In 2003, I received the German Concrete Architecture Prize for a concert hall that I designed inside a gothic red brick cathedral, again using fairface concrete surfaces. My commitment to this material stemmed from the fact that I regarded it to be comparable to natural materials, to be moulded by an architect like a sculptor, or in the same way that Antoni Gaudi shaped natural stone.

No material has been entirely banned in the pursuit of sustainable development, as materials have been listed in order of preference.

The increasing use of wood could perhaps be identified as a megatrend, as escological considerations drive architecture to seek new forms of expression. In some countries, the capability to build even high-rise wood buildings has been boosted by research and the revision of fire safety legislation. Unfortunately, a devastating fire that ravaged a historic wood village in Norway at the time of writing may have a negative impact on the immediate future of wood building. Construction of new buildings, however, is guided by modern town planning, developing materials and updated safety regulations.

It is clear that in the future, material selection will increasingly be guided by how well it supports the goals of sustainable development. One example is green wall with vertical planting covering the whole facade.

Pekka Salminen, Professor, Architect SAFA

PES-Architects

Contents

Selgascano Architecture Office

Taiyuan Museum of Art

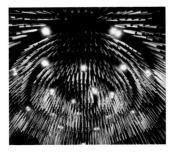

Chalachol Hair Salon

La Estancia Chapel

希尔咖斯卡诺建筑办公室

Selgascano Architecture Office

What is being sought with this studio is quite simple: work under the trees. To do so, we need a roof that is as transparent as possible. Also, at the same time, we need to isolate the desk zone from direct sunlight.

Hence the transparent northern part is covered with a bent sheet of 20 m² colourless plexiglass on the north side. The south side, where the desks are, has to be closed in much more, but not completely, so there is double sheet of fibreglass and polyester in its natural colour on the south side, with translucent insulation in the middle. All three form a 110 mm thick sandwich.

In the former case, the outward view is clear and transparent. The views in the latter case are translucent, somewhat marred by the cantilevered metal structure left inside the sandwich, with the shadow of the trees projecting onto it gently.

Half burying the whole thing, to provide horizontal views of the allotment where the arm is installed, comes before all of that, but it's OK to do it afterwards as well. Everything placed below ground level is in concrete with wood formwork, wooden planks that are also used for paving, firmly bolted, and painted in two colours with two-component paint with an epoxy base.

And to finish off, we have given it a sightly less... slightly more... wet touch: on rainy days, that rain, when it rains, the raindrops on the plastic, when they hit, sometimes more, sometimes less, sometimes a lot... sometimes a sound...

这个工作室所追求的很简单：在树下工作。为了实现这个目标，我们需要一个尽可能透明的屋顶，同时，还要将办公区跟阳光隔离，避免阳光直射。

因此透明的北部覆盖着 20 ㎡ 的无色有机玻璃弧形板。南侧，办公桌所在地，要围起来更多但不是全部都围起来，所以南面是双层自然色的玻璃纤维和聚酯，中间是半透明绝缘体，三者形成一个 110mm 厚的夹层。

前者从外面看是清晰透明的，后者是半透明的，稍微被悬臂金属结构损坏，留在夹层，树木的阴影轻轻地投射到夹层里。

将建筑的一半埋在地面下，首先要考虑为扶手区域营造水平视野，不过这一点在日后完善也无妨。地面下的所有东西都是在混凝土和木模版中，厚木板也用于铺砌地面，被栓紧并涂以两种颜色的双组分环氧基涂料。

最后，我们为建筑添上了或多或少的"湿"意：在下雨天，雨滴轻轻击打着塑料表面，那声音，时而清脆，时而微弱，时而响亮，变幻无常。

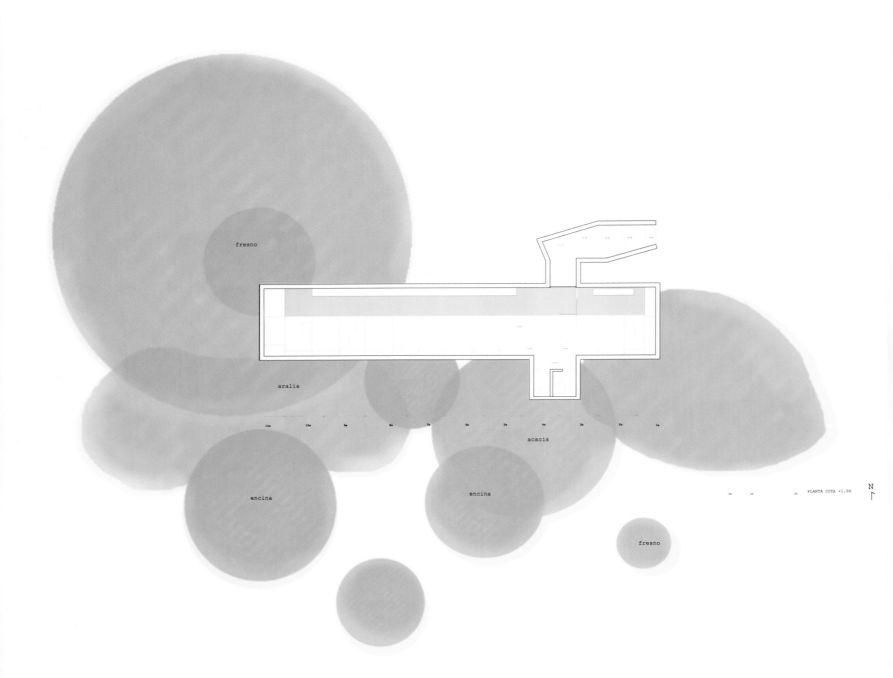

fresno

aralia

acacia

encina

encina

fresno

PLANTA COTA +1.00

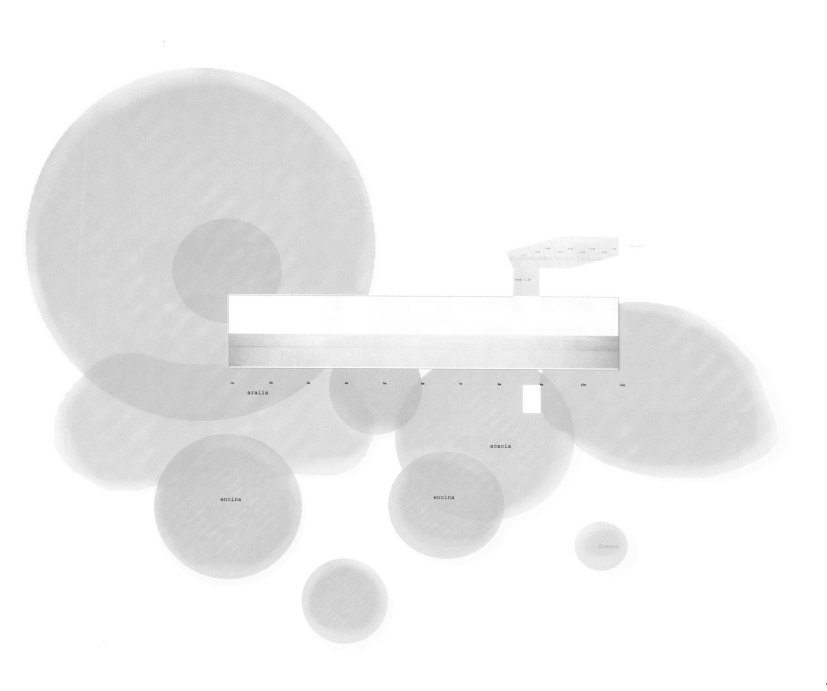

aralia

acacia

encina

encina

fresno

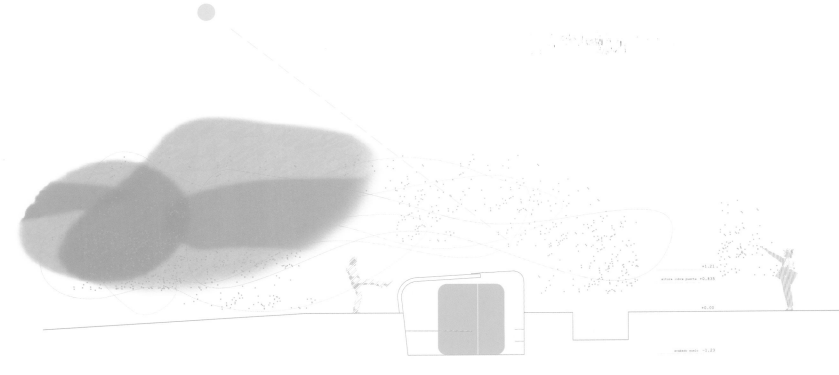

sección A-A'

SECCIÓN POR PUERTA METACRILATO

0m   1m   3m

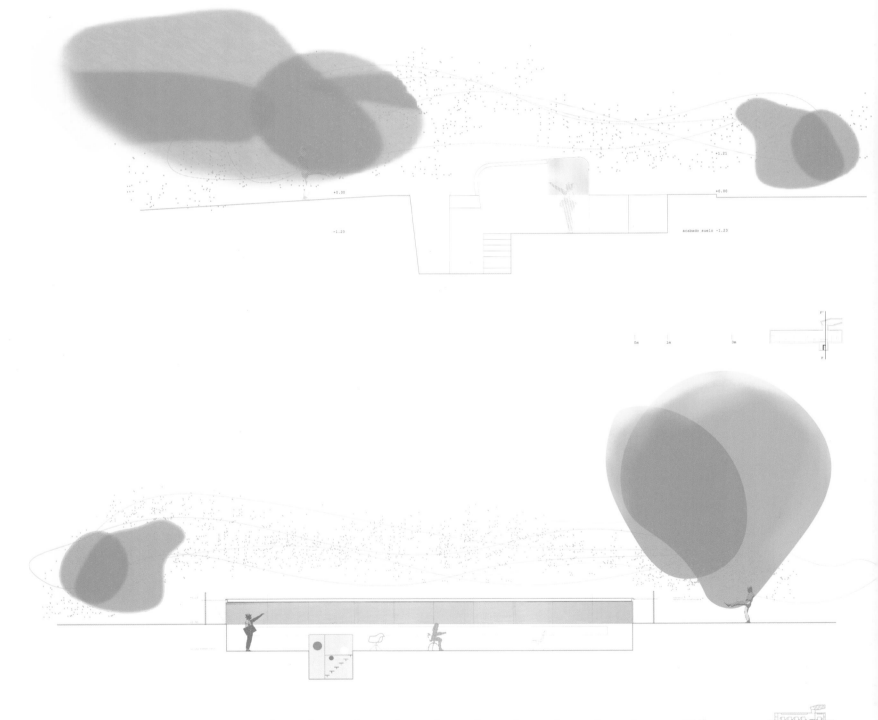

明斯特金色微光展厅
Münster's Glittering Gold Pavilions

The Münster-based architectural group "modulorbeat", consisting of Marc Günnewig and Jan Kampshoff, was commissioned to design the pavilion. modulorbeat developed their new creation for "Golden Glory" together with students from the Münster School of Architecture (msa).

In order to compensate for the slope of the Domplatz, the pavilion was built on top of simple piles made of in-situ concrete and wood, which would leave no traces once they were dismantled. The building has a cross-shaped layout with four "wings". Each wing opens inwards like a funnel, so that the layout is more reminiscent of an abstract windmill than a crucifix. The effect is an inviting gesture to visitors, as the inside path opens up onto the central workshop area. The walls are made of full wooden plywood panels. Indoors, the wood is visible, lending the foundry area a robust, resilient atmosphere. From the outside, the compact pavilion seems similarly robust, but here the facade is covered in eye-catching, gold-copper metal panels. The furrowed, vertical profile of the shimmering panels along each of the four wings narrows towards the centre. This creates a dynamic rhythm, in which the sparkling metal mixes with a play of light and shadows that shifts with the passing of the sun overhead. The small pavilion seems to be a portable, accordion-like construction that could be folded up and transported to another location at any time.

位于明斯特的建筑工作室 modulorbeat（成员为建筑师 Marc Günnewig 和 Jan Kampshoff）被委托设计这一展厅。来自明斯特建筑学校的学生也参与了这个"金色荣光"展厅的创作。

为了适应 Domplatz 地区的坡度，这个展厅被建在直接取材于现场的混凝土和木材简单堆放而成的平台之上，这使得这个临时展厅被拆除后将不会留下任何痕迹。这个建筑呈十字型，有着四个"翼"。每个翼向内开放形成一个漏斗，因此在造型上它更像一个充满怀旧气息的风车而非十字架。建筑内部的通道都能通往位于中心的工作区，充分表达出对游客的邀请。墙体由全木质胶合面板制成。在室内，这些木材肉眼可见，使得整个空间形成一种粗粝而充满弹性的氛围。从外面看，这个盒状展厅有着类似的粗粝感，但外立面覆盖有一层夺人眼球的金铜色金属面板。这些带有沟痕并微微发亮的垂直板紧贴着四翼，并朝着中心的方向越来越窄。这种设计产生了一种非常动态的韵律，金属的光泽与光影的变换随着太阳在头顶的移动而不断改变着。这个小展厅看着就像一个便携的、手风琴似的建筑，似乎能够随时折叠起来移到另一个地方。

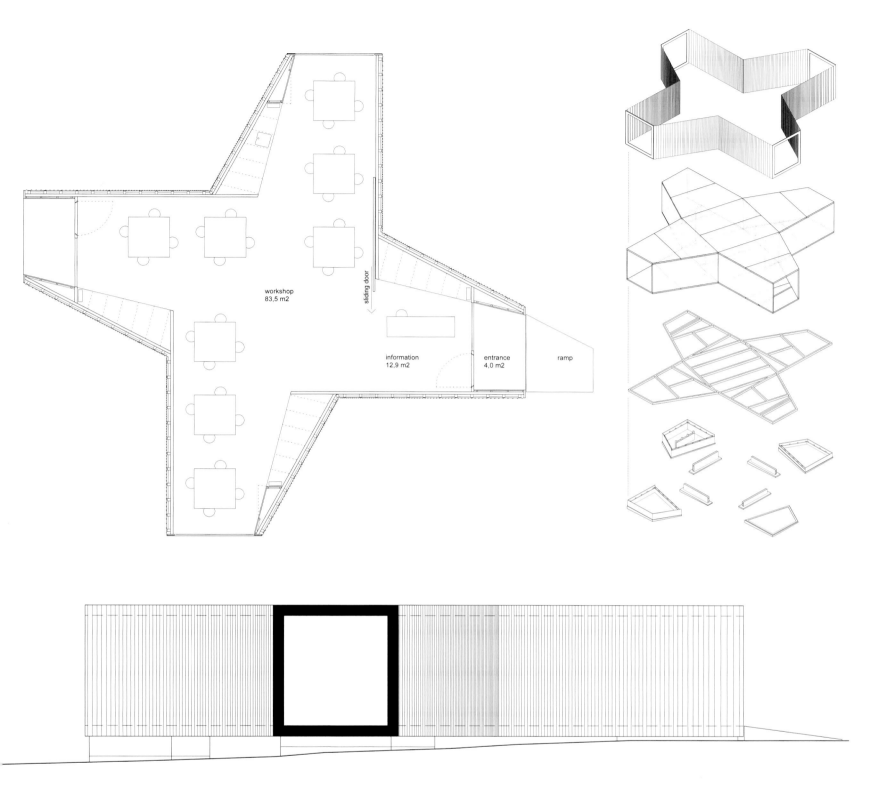

workshop
83,5 m2

sliding door

information
12,9 m2

entrance
4,0 m2

ramp

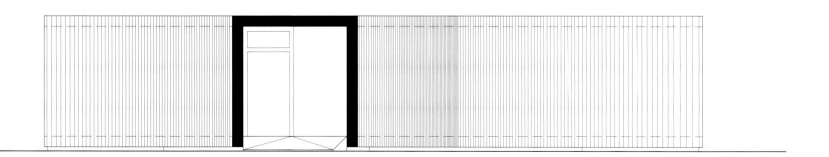

S/W

N/E

Using a very reduced selection of materials and colours, the pavilion's interior design allows for the concentration required by this fine and detailed work. Floor, ceiling and all walls are, like all furniture and other components, crafted from a light coloured wood. This includes the sliding door, through which the foundry can be separated from the information area. In contrast, everything having directly to do with the work has been painted pitch black: the table lamps, the work surfaces, the screw clamps, and even the oil radiators, lamp cables and the kitchen sink.

通过极简的材料和色彩选择，这个展厅的室内设计能够将这项细致的工作要求进行整合。像所有家具和其他设备一样，地板、屋顶和所有的墙面都统一由浅色木材精心制作而成。这包括将工作区与信息区进行分隔的滑动门。极具对比性的是，那些与工作直接相关的东西被涂上了深黑色，例如台灯、工作台、螺丝夹，甚至机油散热器、电灯线和厨房的水槽。

荷兰水运当局办公大楼

Office Building Rijkswaterstaat

The ambition of the client and the architect was to create a sustainable building that explicitly shows the identity of Rijkswaterstaat. The concept was developed with the idea to represent their three core activities (build and maintain highways, waterways and nature) and also integrate the typical ingredients of the site.

The south facade of the building is brutally solid. Through its mass and its horizontal openings, the wall will block the heat of the sun and at the same time stop the noise of the next door highway. The facade, made of concrete and asphalt with lines and patterns, is an abstract representation of a highway. As the rainwater is directed from the roof over the facade, in time, the facade will become green as moss and plants will start growing. This way all 3 aspects of Rijkswaterstaat (road, water and green) are represented in this facade.

At the north facade a light wooden construction creates maximum flexibility and

openness. This facade and its wooden beams form the base for the open layout of the office and the central void where employees of Rijkswaterstaat can meet and exchange knowledge. The green roof terrace also provides the possibility to gather.

The design team looked for the best integration of all disciplines. All technical facilities are integrated into the prefabricated wooden walls and floors. This way an effective building time as well as a maximum of space is realized. The wooden floor will be directly used as a ceiling and the air will flow through the ClimaLevel System directly into the offices. With the Thermal Energy Storage, solar panels on the roof and the use of FCS certified wood, the building gets a class A level in the GreenCalc calculation method and could also be classified as a BREEAM excellent building.

客户和建筑师的目标是创造一个可持续发展且明确代表 Rijkswaterstaat 身份的建筑。设计理念来自代表其三个核心业务（修建和维护高速公路、航道和大自然）的想法，同时也融入到当地的典型环境中。

建筑的南立面是无比坚固的实心体。通过建筑块和水平开口，墙壁可阻隔太阳的热量，同时也可屏蔽隔壁高速公路的噪声。建筑外观由带有线条、图案的混凝土和沥青构成，是高速公路的抽象表现。当雨水流过屋顶的外层时，经过一段时间后，绿色苔藓植物将开始生长。这样，Rijkswaterstaat 的三个业务（道路、水路和绿化）都在这个建筑外观中表现出来了。

在建筑的北立面，轻型木制结构创造了最大的灵活性和开放性。这个外层和木梁形成了地基，作为办公室的开放式布局和中央空地，在那里 Rijkswaterstaat 的员工可以见面和交流。绿色屋顶平台还提供了聚会的可能性。

设计团队在所有学科中寻找最好的整合，所有的科技设备都整合到预制的木墙和地板上。这样一来，一个有效的时间以及空间的最大化就实现了。木地板将直接作为天花板，空气经通风系统直接进入办公室。FSC 认证木材的使用，与屋顶的储热器、太阳能电池板一起，让该建筑获得了 A 级 GreenCalc 评估认证，也被评为 BREEAM 优秀级建筑。

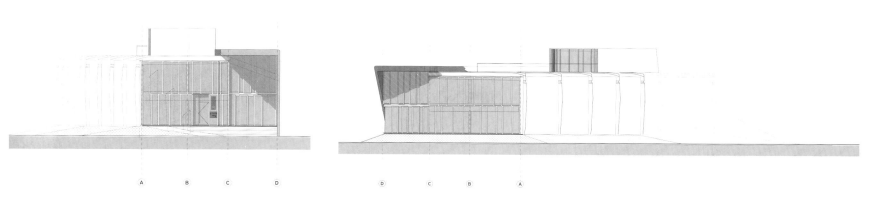

west facade

east facade

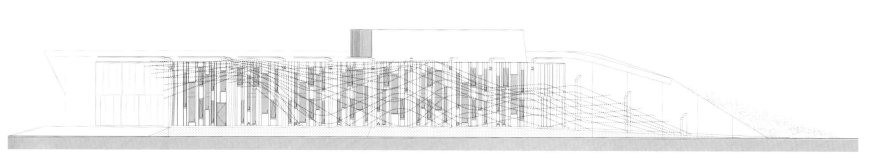

north facade

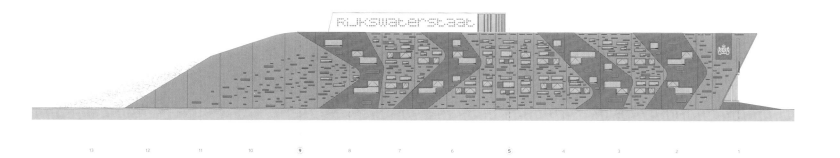

south facade

concept model

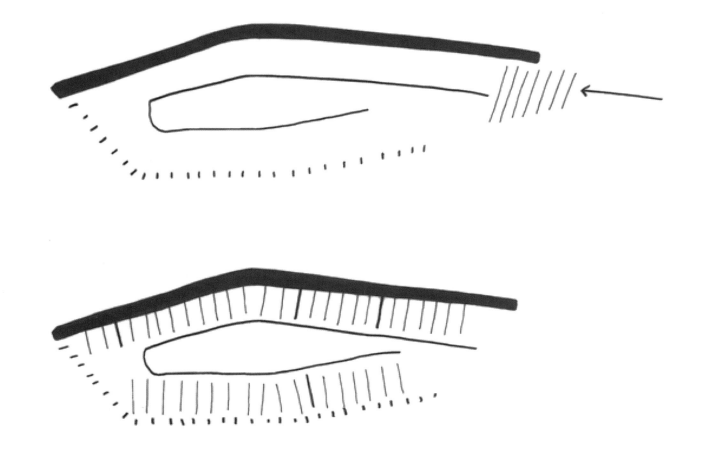

sketch flexibility

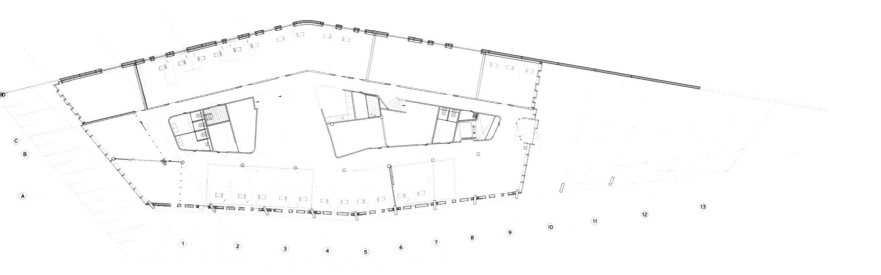

ground floor

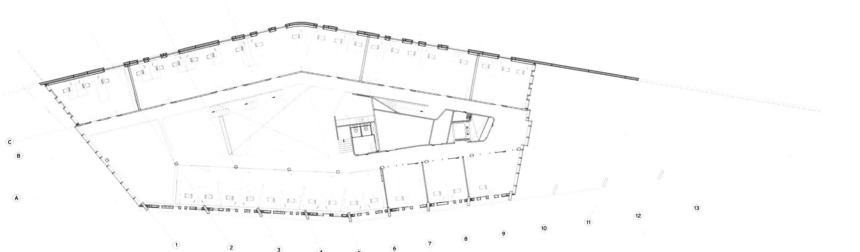

first floor

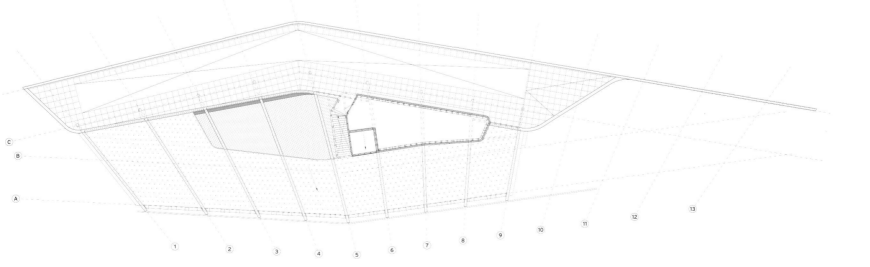

roof

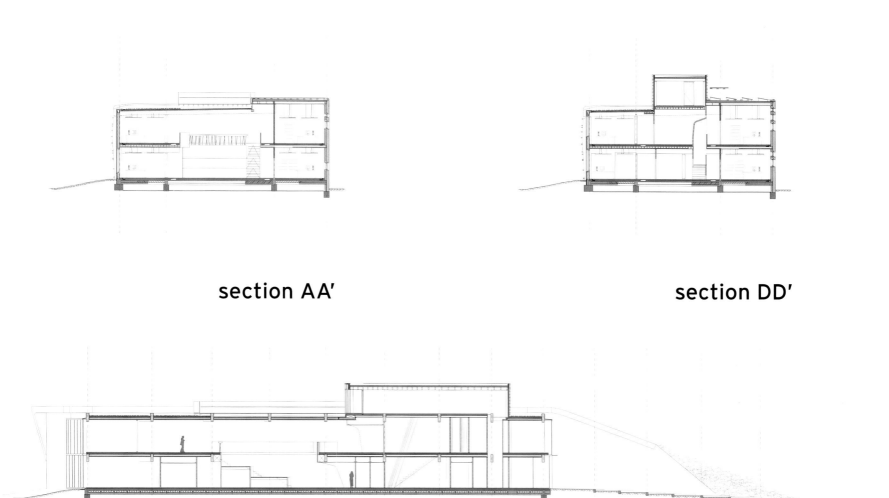

section AA'

section DD'

section BB'

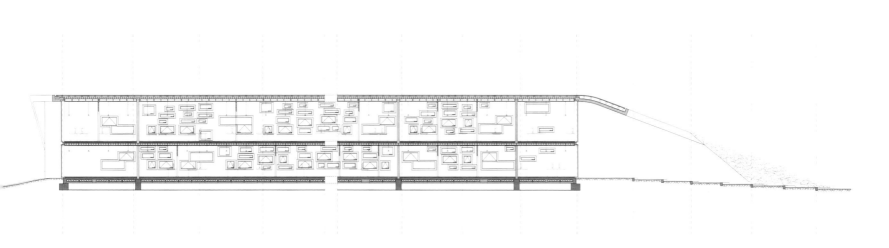

section CC'

滑铁卢博物馆
Waterloo Region Museum

The Waterloo Region Museum is located on the beautiful and pastoral grounds of Doon Heritage Village — a living history village that has interpreted regional history since 1957. The Museum acts as a visitor centre for the living history village, also contains more than 1858 m² of gallery space in order to tell a much more complete history of the region, its history and its people. The building contains a large central lobby space, community and classroom spaces for school and conference programs, a 120-seat theatre, and a dramatic "Quilt Wall" facade facing the main street. This LEED Silver project is the first, and currently the only museum in Canada that has achieved a LEED Certification.

　　滑铁卢博物馆位于美丽的田园乡村——杜恩遗产村，这是一座自1957年起就阐释着这片区域历史的现存历史乡村。博物馆作为一个历史村的游客中心，包含了超过1858 m²的画廊空间，以讲述更完整的区域历史、馆史和人民。这栋建筑包含一个宽敞的中央大厅，学校与会议的社区和教室，一个120座的剧院和显眼的正对着主街道的"棉被墙"外观。这是首个也是目前唯一一个在加拿大获得LEED银级资格认证的博物馆。

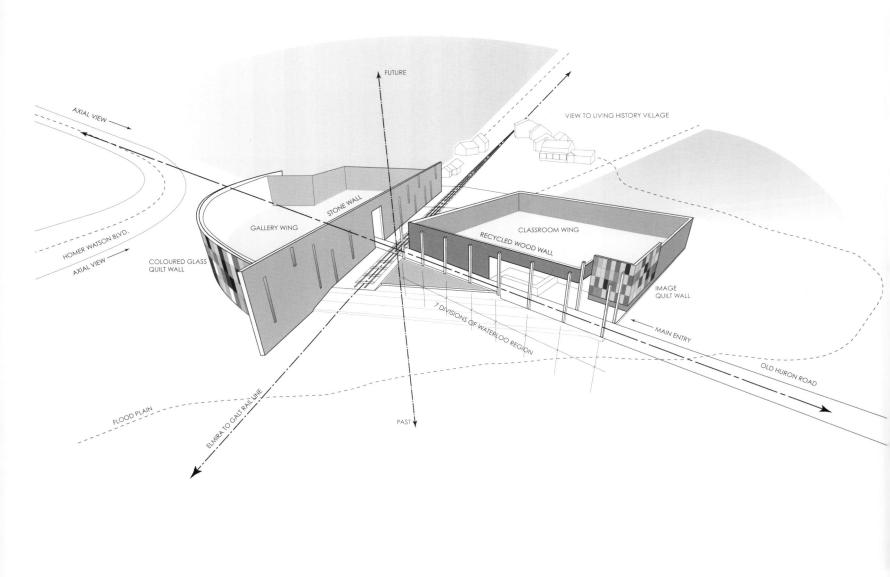

AXIAL VIEW

FUTURE

VIEW TO LIVING HISTORY VILLAGE

HOMER WATSON BLVD.

AXIAL VIEW

STONE WALL

GALLERY WING

CLASSROOM WING

COLOURED GLASS
QUILT WALL

RECYCLED WOOD WALL

IMAGE
QUILT WALL

7 DIVISIONS OF WATERLOO REGION

MAIN ENTRY

OLD HURON ROAD

FLOOD PLAIN

ELMIRA TO GALT RAIL LINE

PAST

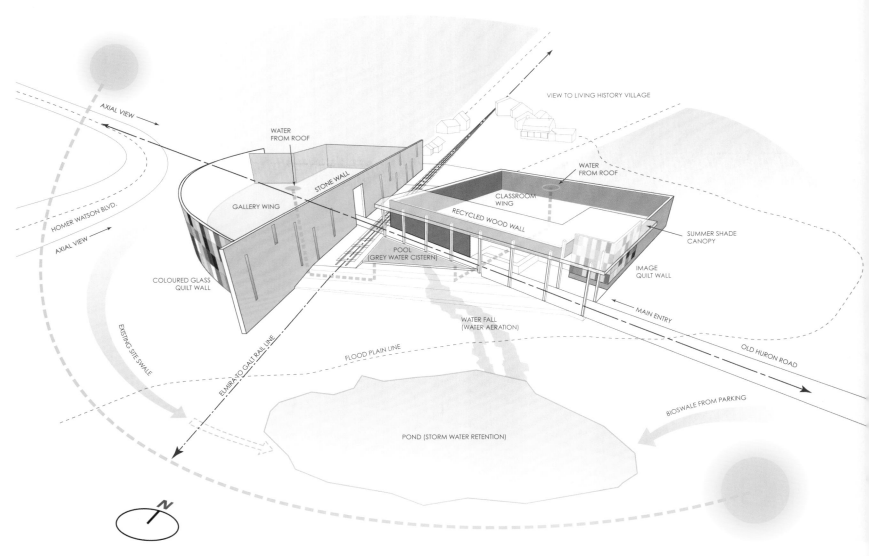

AXIAL VIEW

VIEW TO LIVING HISTORY VILLAGE

WATER
FROM ROOF

WATER
FROM ROOF

HOMER WATSON BLVD.

STONE WALL

GALLERY WING

CLASSROOM
WING

AXIAL VIEW

RECYCLED WOOD WALL

SUMMER SHADE
CANOPY

COLOURED GLASS
QUILT WALL

POOL
(GREY WATER CISTERN)

IMAGE
QUILT WALL

EXISTING SITE SWALE

ELMIRA TO GALT RAIL LINE

WATER FALL
(WATER AERATION)

MAIN ENTRY

OLD HURON ROAD

FLOOD PLAIN LINE

BIOSWALE FROM PARKING

POND (STORM WATER RETENTION)

N

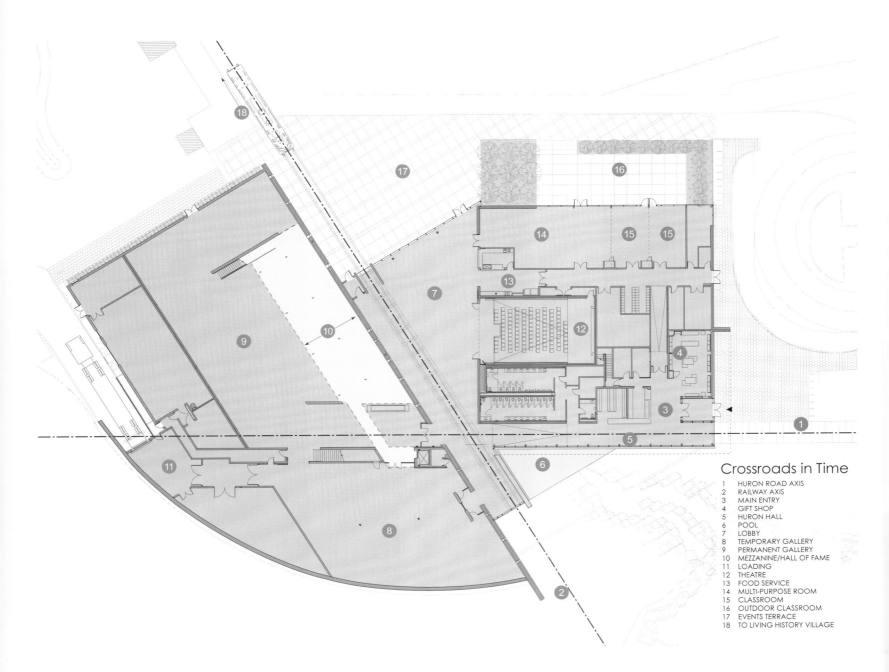

Crossroads in Time

1 HURON ROAD AXIS
2 RAILWAY AXIS
3 MAIN ENTRY
4 GIFT SHOP
5 HURON HALL
6 POOL
7 LOBBY
8 TEMPORARY GALLERY
9 PERMANENT GALLERY
10 MEZZANINE/HALL OF FAME
11 LOADING
12 THEATRE
13 FOOD SERVICE
14 MULTI-PURPOSE ROOM
15 CLASSROOM
16 OUTDOOR CLASSROOM
17 EVENTS TERRACE
18 TO LIVING HISTORY VILLAGE

The primary sustainable features of the project are:
- The interpretation of water usage by means of a storm water pond functioning as a grey water reservoir;
- Extensive use of highly visible recycled materials of significant local historic significance;
- A priority placed on sourcing regional materials to be used in its construction;
- The enhancement of the local ecology and history of the site through building placement and design;
- Education of the public about in their local history as well as specific sustainability objectives.

项目的主要可持续发展特色有：
- 通过作为废水储水池的雨水池诠释水的使用；
- 有当地重要历史意义，可高效循环再用材料的广泛应用；
- 施工优先就地取材；
- 通过建筑安置和设计，增强了当地的生态和历史；
- 有关当地历史和特定的可持续性目标的公共教育

SOUTH ELEVATION

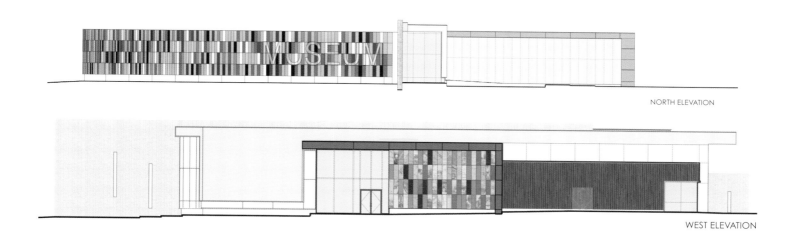

NORTH ELEVATION

WEST ELEVATION

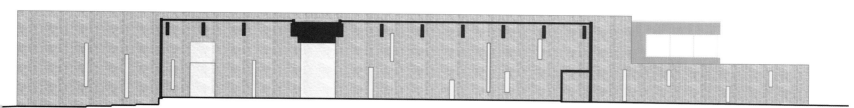

RAILWAY SECTION

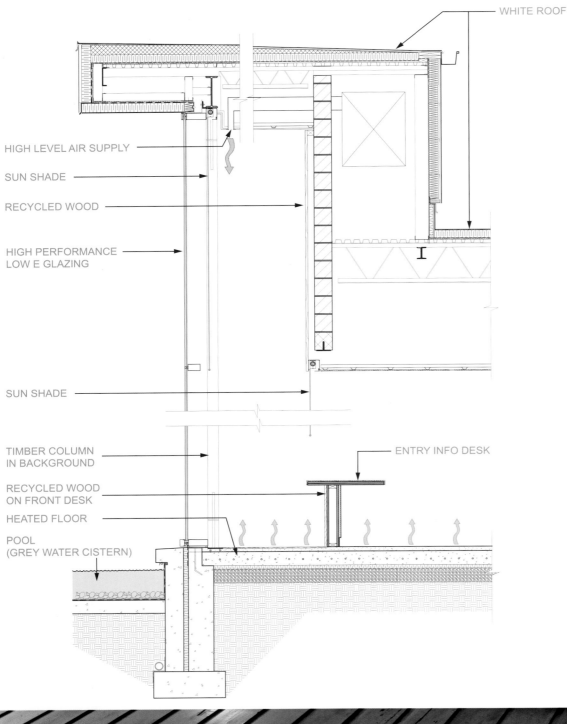

WHITE ROOF

HIGH LEVEL AIR SUPPLY

SUN SHADE

RECYCLED WOOD

HIGH PERFORMANCE
LOW E GLAZING

SUN SHADE

TIMBER COLUMN
IN BACKGROUND

ENTRY INFO DESK

RECYCLED WOOD
ON FRONT DESK

HEATED FLOOR

POOL
(GREY WATER CISTERN)

海洋冲浪博物馆

Cité de l'Océan et du Surf

The Cité de l'Océan et du Surf is a museum that explores both surf and sea and their role upon our leisure, science and ecology. The building form derives from the spatial concept "under the sky" / "under the sea". A concave "under the sky" shape forms the character of the main exterior plaza, the "Place de l'Océan", which is open to the sky and sea, with the horizon in the distance. A convex structural ceiling forms the "under the sea" exhibition spaces.

Two "glass boulders," which contain the restaurant and the surfer's kiosk, activate the central outdoor plaza and connect analogically to the two great boulders on the beach in the distance. The building's southwest corner is dedicated to the surfers' hangout with a skate pool at the top and an open porch underneath that connects to the auditorium and exhibition spaces inside the museum. This covered area provides a sheltered space for outdoor interaction, meetings and events.

海洋冲浪博物馆是一个可探索冲浪与海洋及其对人类休闲、科学和生态的作用的博物馆。其建筑形状源于空间概念"天空下"和"海洋里"。"天空下"的凹形构成了"海洋广场"这个主要户外广场的特点，这个广场与远处的地平线一起，通向天空和海洋。凸形结构的天花板形成了"海洋里"的展览空间。

如两块"玻璃巨石"的建筑设有餐厅和冲浪者的休息棚，可为中央户外广场带来活力并连接远处沙滩上类似的两大巨石。大楼的西南角，顶部是专为冲浪者设计的带有溜冰场的游乐场，下面是连接博物馆礼堂和展馆的开放门廊。这个设有顶篷的区域，为户外互动、会议和活动提供了一个遮风避雨的空间。

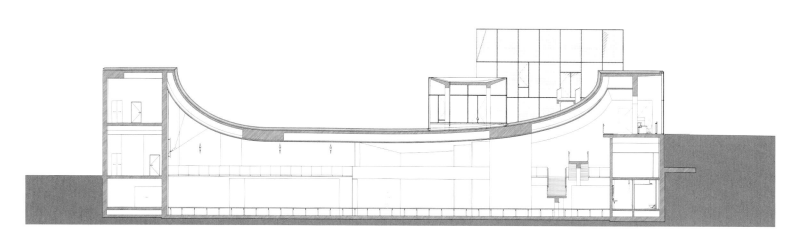

CROSS SECTION

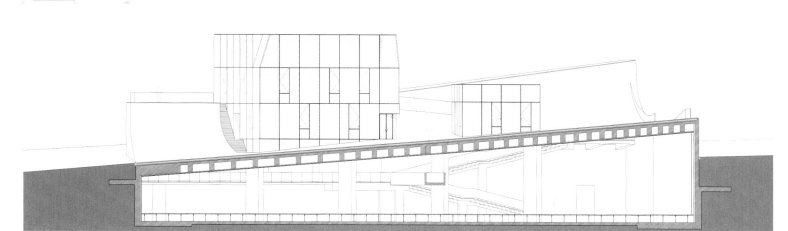

LONG SECTION

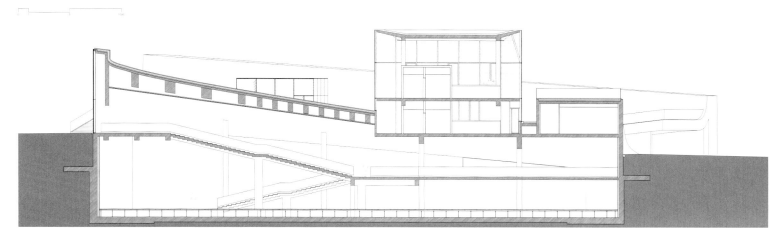

LOBBY SECTION

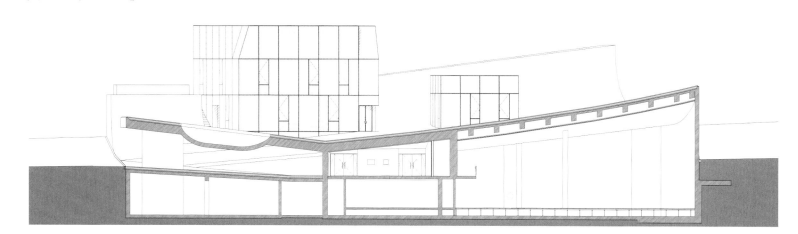

PORCH SECTION

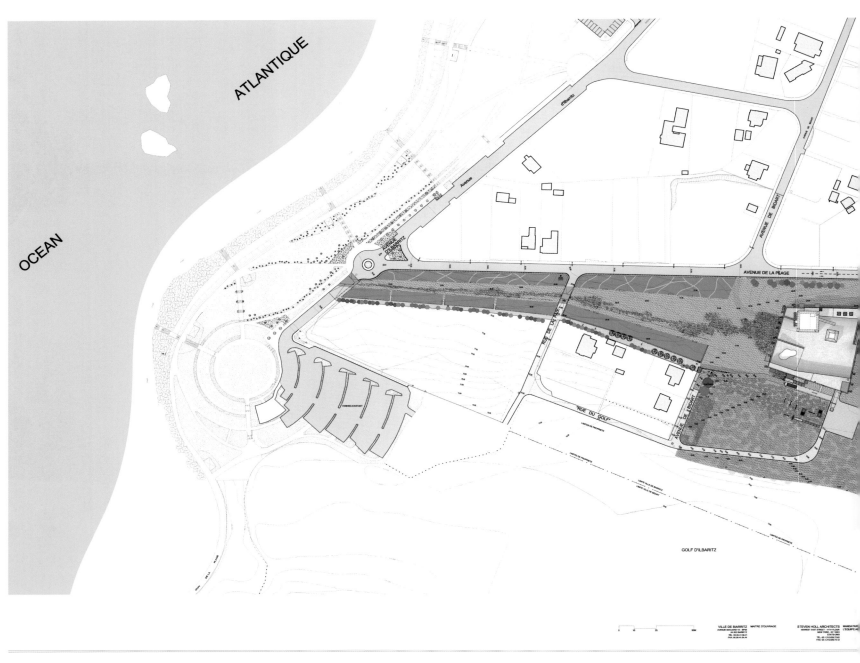

OCEAN

ATLANTIQUE

GOLF D'ILBARITZ

VILLE DE BIARRITZ   MAITRE D'OUVRAGE
AVENUE EDOUARD VII - BP96
64 202 BIARRITZ
TEL 05.59.41.59.41
FAX 05.59.41.59.41

STEVEN HOLL ARCHITECTS   MANDATAIRE
435 WEST 31ST STREET - 11TH FLOOR   L'EQUIPE AU
NEW YORK - NY 10001
ETATS-UNIS
TEL 00 1 212.629.70.60
FAX 00 1 212.629.70.18

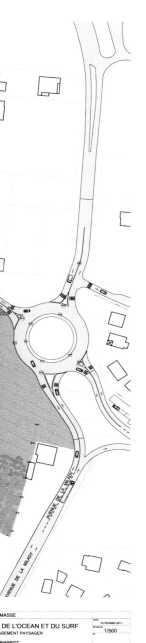

MASSE

DE L'OCEAN ET DU SURF

AGEMENT PAYSAGER

BIARRITZ

22 FEVRIER 2011

1/500

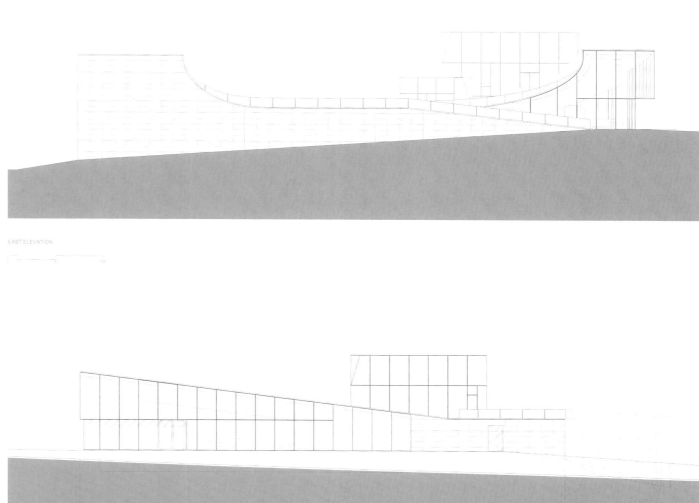

EAST ELEVATION

NORTH ELEVATION

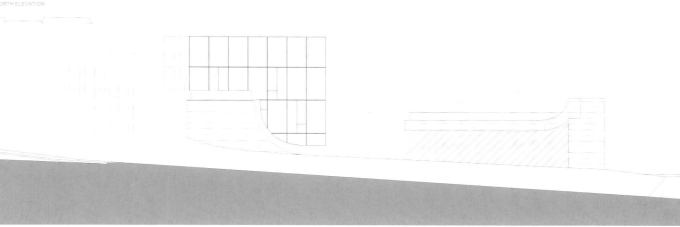

WEST ELEVATION

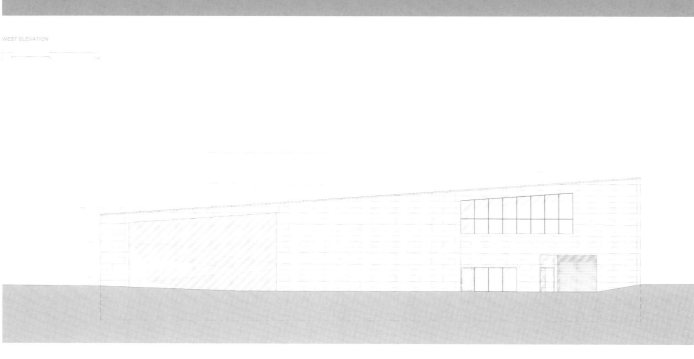

SOUTH ELEVATION

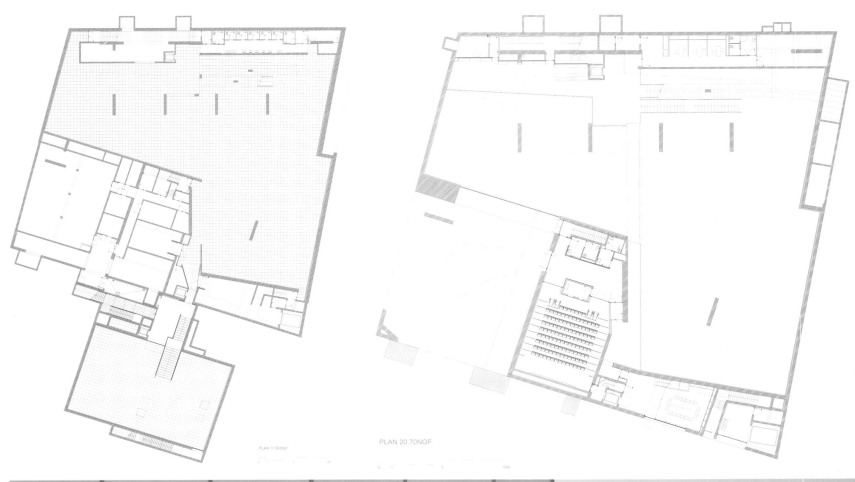

PLAN 17.50NGF   PLAN 20.70NGF

The gardens of the Cité de l'Océan et du Surf aim at a fusion of architecture and landscape, and connect the project to the ocean horizon. The precise integration of concept and topography gives the building its unique profile. The materials of the public plaza are a progressive variation of Portuguese cobblestone paving with grass and natural vegetation. Towards the ocean, the concave form of the building plaza is extended through the landscape. With slightly cupped edges, these gardens, a mix of field and local vegetation, are a continuation of the building and will host festivals and daily events that are integrated with the museum facilities.

海洋冲浪博物馆的花园意将建筑与景观融为一体，并连接项目本身与海平面。设计理念和地形的精确结合让建筑拥有了独特的轮廓。公共广场是葡式鹅卵石与草和自然植被一起铺设的渐进变化的路面。面向海洋，建筑广场的凹形通过景观得以延伸。与稍微凹陷的边缘一起，这些混合了场所和当地植被的花园，是建筑的延续，将举办与博物馆设施融为一体的节日聚会和日常活动。

纳尔逊－阿特金斯艺术博物馆

Nelson-Atkins Museum of Art

The Bloch Building, a new addition to the Nelson-Atkins Museum of Art, fuses architecture with landscape to engage the existing sculpture garden. The new building is distinguished by five "glass lenses", traversing from the existing building through the Sculpture Park to form new spaces and angles of vision.

The new elements exist in complementary contrast with the original 1933 classical museum building. The first of the five "lenses" forms a bright and transparent lobby, with café, art library and bookstore, inviting the public into the Museum and encouraging movement via ramps toward the galleries as they progress downward into the garden. From the lobby a new cross-axis connects through to the original building's grand spaces.

布洛赫大楼是纳尔逊－阿特金斯艺术博物馆的新建筑，它将建筑和景观融入到现有的雕塑花园中。新建筑以五块"玻璃透镜"闻名，横穿过雕塑公园的现有建筑，形成新的空间和视角。

新建筑与创建于 1933 的古典博物馆建筑形成互补。五块"透镜"中的第一块形成一条明亮透明的带有咖啡厅、艺术图书馆和书店的走廊，它可引导公众进入博物馆，并在他们下行走进花园时，鼓励他们通过弯道进入博物馆。从大堂开始，一个新的十字轴连接到原建筑的高大空间。

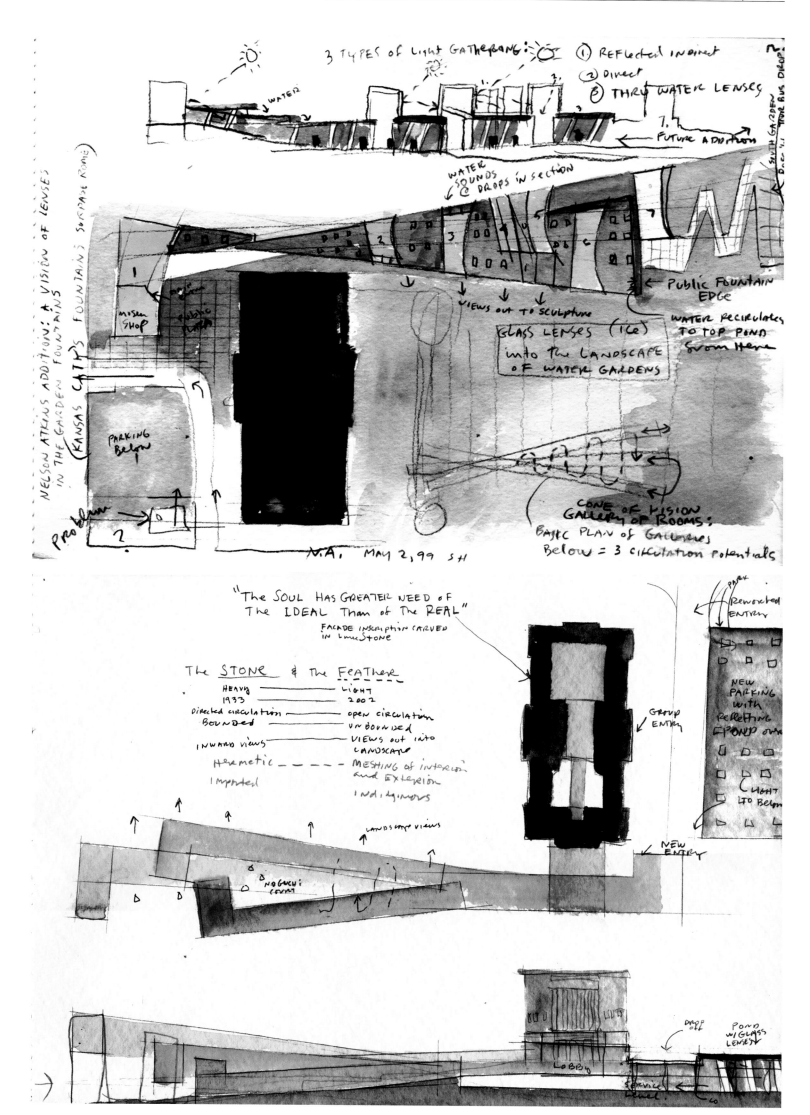

3 TYPES OF LIGHT GATHERING: ① REFLECTED INDIRECT
② DIRECT
③ THRU WATER LENSES

WATER

FUTURE ADDITION

NELSON ATKINS ADDITION: A VISION OF LENSES IN THE GARDEN FOUNTAINS
(KANSAS CATH'S FOUNTAINS SURFACE POND)

WATER SOUNDS @ DROPS IN SECTION

PUBLIC FOUNTAIN EDGE

MUSEUM SHOP

PUBLIC PLAZA

VIEWS OUT TO SCULPTURE

GLASS LENSES (ICE) into the LANDSCAPE OF WATER GARDENS

WATER RECIRCULATES TO TOP POND FROM HERE

PARKING Below

PROBLEM

CONE OF VISION GALLERY OF ROOMS: BASIC PLAN OF GALLERIES Below = 3 CIRCULATION potentials

N.A. MAY 2, 99 SH

"The SOUL HAS GREATER NEED of THE IDEAL Than of The REAL"

FACADE INSCRIPTION CARVED IN LIMESTONE

The STONE & The FEATHER
HEAVY ——————— LIGHT
1933 ——————— 2002
Directed circulation ——— OPEN CIRCULATION
BOUNDED ——————— UNBOUNDED
INWARD views ——————— VIEWS OUT INTO LANDSCAPE
Hermetic — — — — — MESHING OF INTERIOR and EXTERIOR
Imported   INDIGENOUS

PARK
REWORKED ENTRY

GROUP ENTRY

NEW PARKING with REFLECTING POND OVER

LIGHT LTD Below

NEW ENTRY

LANDSCAPE VIEWS

NOGUCHI COURT

DROP OFF

POND W/GLASS LENSES

LOBBY

SERVICE LEVEL

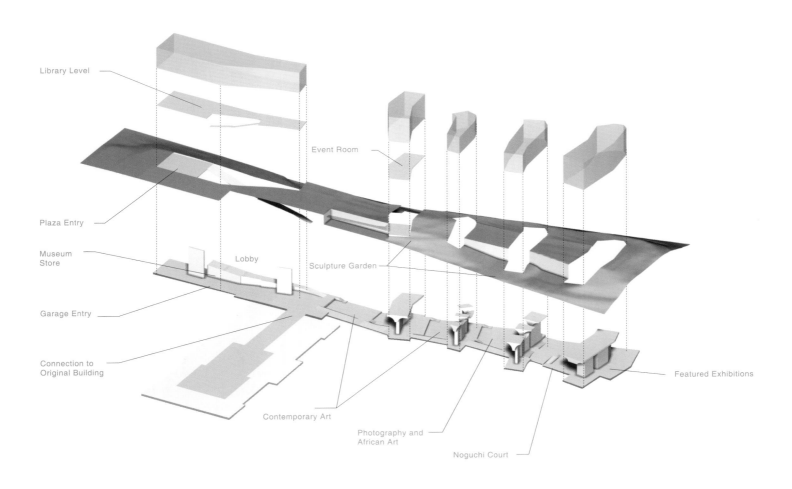

Library Level

Event Room

Plaza Entry

Museum Store

Lobby

Sculpture Garden

Garage Entry

Connection to Original Building

Contemporary Art

Photography and African Art

Noguchi Court

Featured Exhibitions

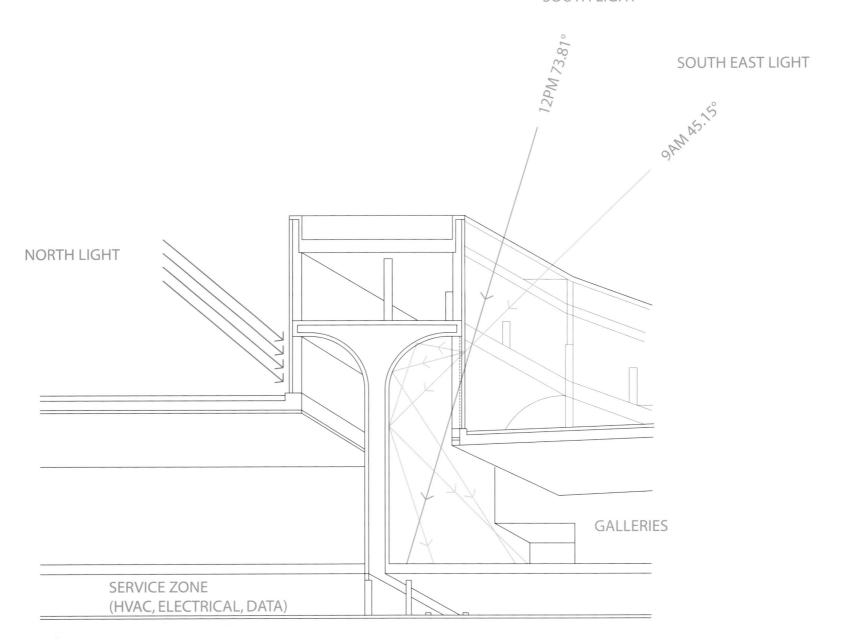

SOUTH LIGHT

SOUTH EAST LIGHT

12PM 73.81°

9AM 45.15°

NORTH LIGHT

GALLERIES

SERVICE ZONE
(HVAC, ELECTRICAL, DATA)

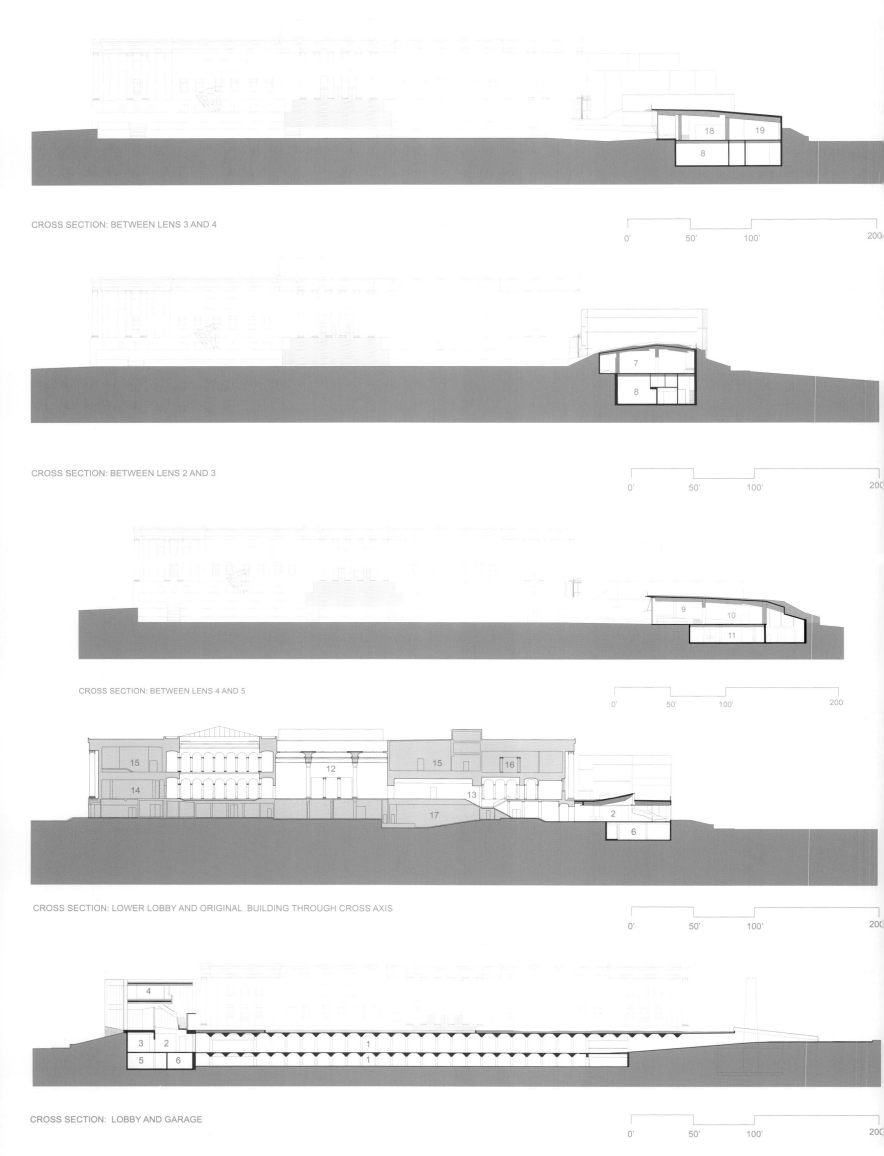

CROSS SECTION: BETWEEN LENS 3 AND 4

0'   50'   100'   200'

CROSS SECTION: BETWEEN LENS 2 AND 3

0'   50'   100'   200'

CROSS SECTION: BETWEEN LENS 4 AND 5

0'   50'   100'   200'

CROSS SECTION: LOWER LOBBY AND ORIGINAL BUILDING THROUGH CROSS AXIS

0'   50'   100'   200'

CROSS SECTION: LOBBY AND GARAGE

0'   50'   100'   200'

Bloch Building

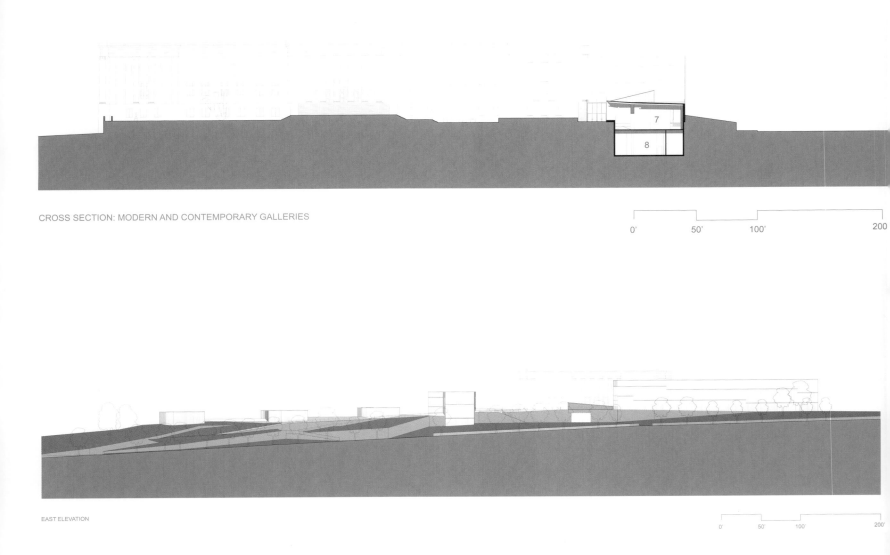

CROSS SECTION: MODERN AND CONTEMPORARY GALLERIES

0'   50'   100'   200

EAST ELEVATION

0'   50'   100'   200'

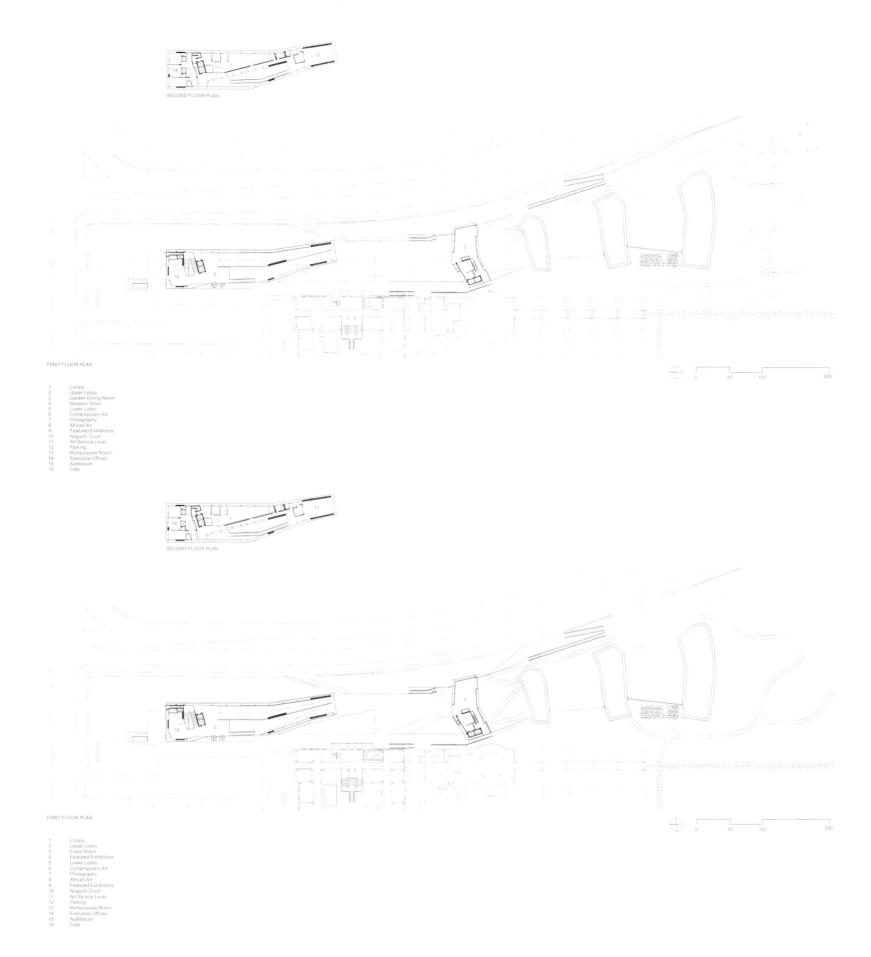

SECOND FLOOR PLAN

FIRST FLOOR PLAN

1 Library
2 Upper Lobby
3 Garden Dining Room
4 Museum Store
5 Lower Lobby
6 Contemporary Art
7 Photography
8 African Art
9 Featured Exhibitions
10 Noguchi Court
11 Art Service Level
12 Parking
13 Multipurpose Room
14 Executive Offices
15 Auditorium
16 Cafe

SECOND FLOOR PLAN

FIRST FLOOR PLAN

1 Library
2 Upper Lobby
3 Event Room
4 Featured Exhibitions
5 Lower Lobby
6 Contemporary Art
7 Photography
8 African Art
9 Featured Exhibitions
10 Noguchi Court
11 Art Service Level
12 Parking
13 Multipurpose Room
14 Executive Offices
15 Auditorium
16 Cafe

The lenses' multiple layers of translucent glass gather, diffuse and refract light. During the day the lenses inject varying qualities of light into the galleries, while at night the sculpture garden glows with their internal light.

The design for the new addition employs several sustainable building concepts:

green roofs achieve high insulation and control storm water; The structural concept of the lenses merges with a light and air distributor concept; The double-glass cavities of the lenses gather sun-heated air in winter or exhaust it in summer; Optimum light levels are ensured through the use of computer-controlled screens and of special translucent insulating material embedded in the glass cavities.

透镜的多层半透明玻璃收集、散射和折射着光线,在白天透镜将不同质量的光线折射入画廊,而在夜晚,雕塑花园则利用内部照明发光。

新大楼的设计采用了几种可持续发展建筑的概念:绿色屋顶实现了高效隔热

和雨水控制;透镜的结构概念融合了光和空气分流器的概念;透镜的双层玻璃的中空在冬季收集受太阳加热的空气或在夏季排出热空气;最佳的光照水平通过使用计算机控制的格栅和嵌在玻璃间隙中特殊的半透明隔热材料来保障。

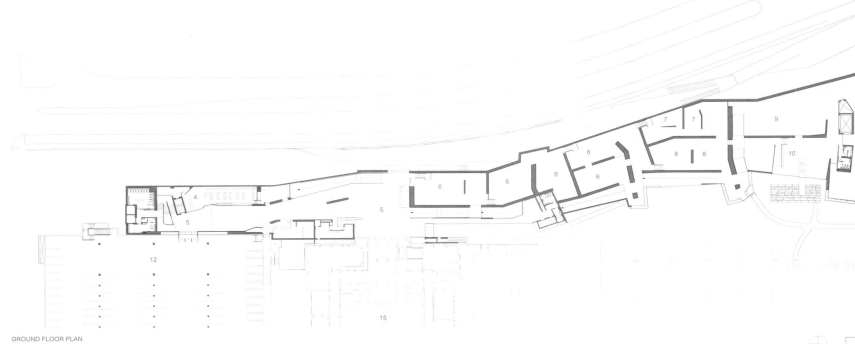

GROUND FLOOR PLAN

1   Library
2   Upper Lobby
3   Event Room
4   Museum Store
5   Lower Lobby
6   Contemporary Art
7   Photography
8   African Art
9   Special Exhibitions
10   Noguchi Court
11   Art Service Level
12   Parking
13   Multipurpose Room
14   Executive Offices
15   Auditorium
16   Cafe

"T" PLAN DIAGRAM

"T" AXON DIAGRAM

"T" PERSPECTIVE DIAGRAM

"T" CROSS SECTION DIAGRAM

LENS 3 "T" WALL

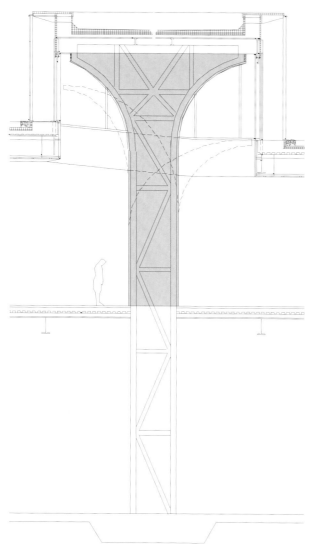

"T" SECTION DIAGRAM LENS 3

200'

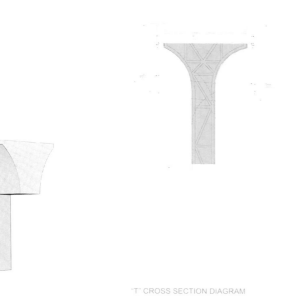

"T" CROSS SECTION DIAGRAM

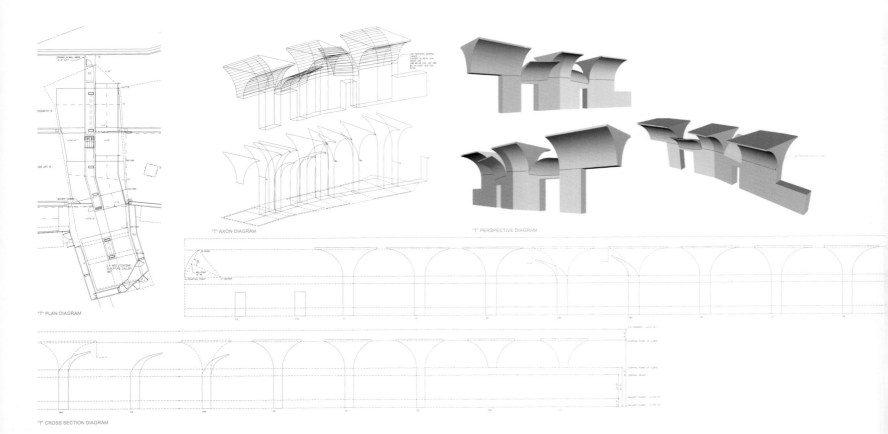

"T" PLAN DIAGRAM

"T" AXON DIAGRAM

"T" PERSPECTIVE DIAGRAM

"T" CROSS SECTION DIAGRAM

LENS 4 "T" WALL

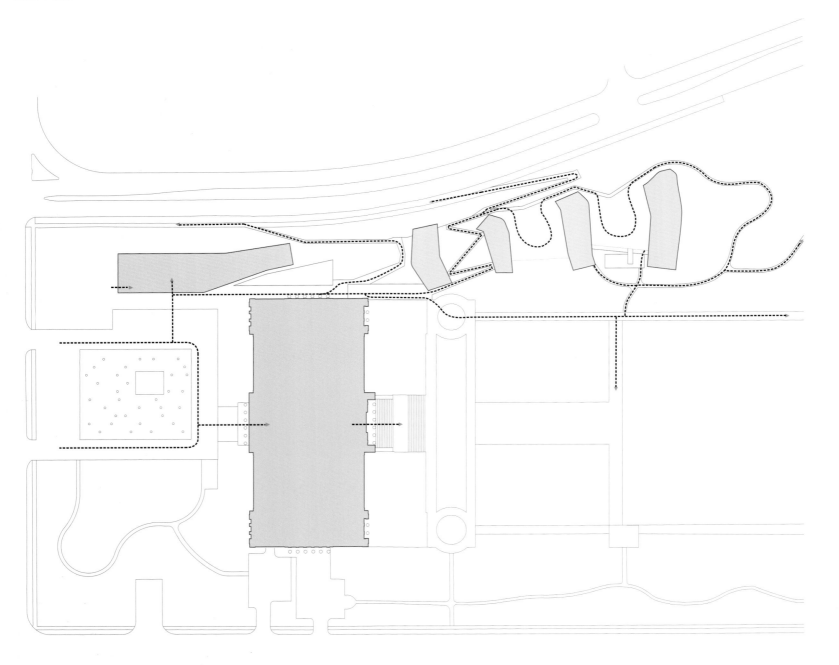

SCULPTURE GARDEN CIRCULATION

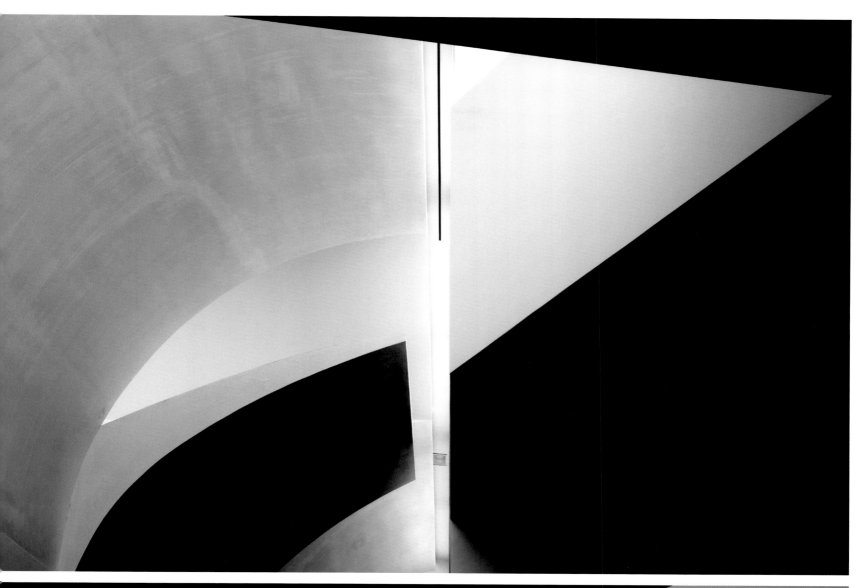

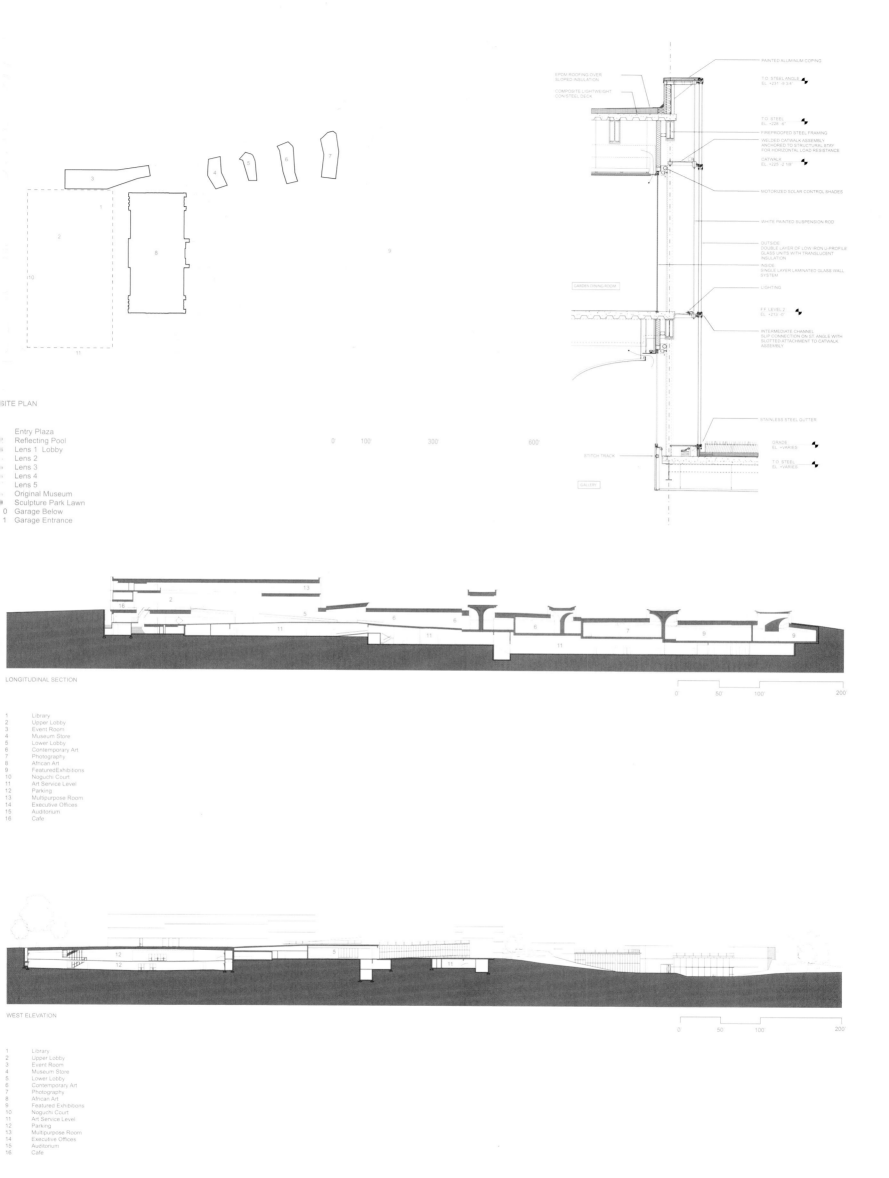

SITE PLAN

Entry Plaza
Reflecting Pool
Lens 1 Lobby
Lens 2
Lens 3
Lens 4
Lens 5
Original Museum
Sculpture Park Lawn
0   Garage Below
1   Garage Entrance

0   100'   300'   600'

PAINTED ALUMINUM COPING

EPDM ROOFING OVER
SLOPED INSULATION

COMPOSITE LIGHTWEIGHT
CON/STEEL DECK

T.O. STEEL ANGLE
EL +231'-9 3/4"

T.O. STEEL
EL +228'-6"

FIREPROOFED STEEL FRAMING

WELDED CATWALK ASSEMBLY
ANCHORED TO STRUCTURAL STAY
FOR HORIZONTAL LOAD RESISTANCE

CATWALK
EL +225'-2 1/8"

MOTORIZED SOLAR CONTROL SHADES

WHITE PAINTED SUSPENSION ROD

OUTSIDE
DOUBLE LAYER OF LOW IRON U-PROFILE
GLASS UNITS WITH TRANSLUCENT
INSULATION

INSIDE
SINGLE LAYER LAMINATED GLASS WALL
SYSTEM

LIGHTING

GARDEN DINING ROOM

F.F. LEVEL 2
EL +213'-0"

INTERMEDIATE CHANNEL
SLIP CONNECTION ON ST. ANGLE WITH
SLOTTED ATTACHMENT TO CATWALK
ASSEMBLY

STAINLESS STEEL GUTTER

STITCH TRACK

GALLERY

GRADE
EL +VARIES

T.O. STEEL
EL +VARIES

LONGITUDINAL SECTION

0   50'   100'   200'

1   Library
2   Upper Lobby
3   Event Room
4   Museum Store
5   Lower Lobby
6   Contemporary Art
7   Photography
8   African Art
9   Featured Exhibitions
10   Noguchi Court
11   Art Service Level
12   Parking
13   Multipurpose Room
14   Executive Offices
15   Auditorium
16   Cafe

WEST ELEVATION

0   50'   100'   200'

1   Library
2   Upper Lobby
3   Event Room
4   Museum Store
5   Lower Lobby
6   Contemporary Art
7   Photography
8   African Art
9   Featured Exhibitions
10   Noguchi Court
11   Art Service Level
12   Parking
13   Multipurpose Room
14   Executive Offices
15   Auditorium
16   Cafe

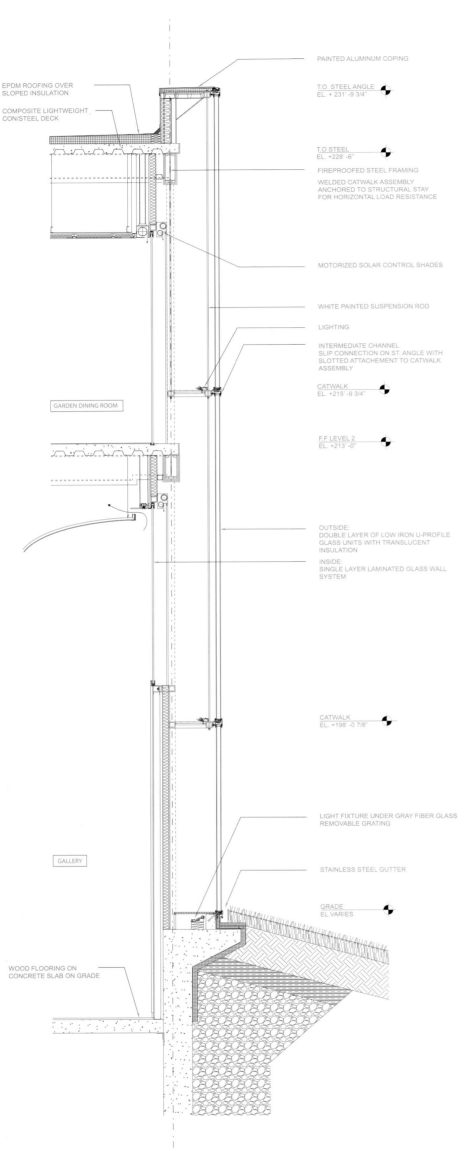

PAINTED ALUMINUM COPING

EPDM ROOFING OVER
SLOPED INSULATION

COMPOSITE LIGHTWEIGHT
CON/STEEL DECK

T.O. STEEL ANGLE
EL. + 231'-9 3/4"

T.O STEEL
EL. +228'-6"

FIREPROOFED STEEL FRAMING

WELDED CATWALK ASSEMBLY
ANCHORED TO STRUCTURAL STAY
FOR HORIZONTAL LOAD RESISTANCE

MOTORIZED SOLAR CONTROL SHADES

WHITE PAINTED SUSPENSION ROD

LIGHTING

INTERMEDIATE CHANNEL
SLIP CONNECTION ON ST. ANGLE WITH
SLOTTED ATTACHEMENT TO CATWALK
ASSEMBLY

CATWALK
EL. +215'-9 3/4"

GARDEN DINING ROOM

F.F LEVEL 2
EL. +213'-0"

OUTSIDE:
DOUBLE LAYER OF LOW IRON U-PROFILE
GLASS UNITS WITH TRANSLUCENT
INSULATION

INSIDE:
SINGLE LAYER LAMINATED GLASS WALL
SYSTEM

CATWALK
EL. +198'-0 7/8"

LIGHT FIXTURE UNDER GRAY FIBER GLASS
REMOVABLE GRATING

GALLERY

STAINLESS STEEL GUTTER

GRADE
EL.VARIES

WOOD FLOORING ON
CONCRETE SLAB ON GRADE

太原美术馆
Taiyuan Museum of Art

The Taiyuan Museum of Art works as a cluster of buildings unified by continuous and discontinuous promenades both inside and outside. The building responds to the urban parkscape in which it is set, visitors are encouraged to pass through the building while not entering into the museum itself. An exterior ramp threading through the building connects the heterogeneous hardscapes, lawns and sculpture gardens. The integration of building and landscape registers multiple scales of territory ranging from the enormity of the adjacent Fen River to the intimacy of the museum's own particular spatial episodes.

Inside, the security of museum space is maintained by a highly controlled interface between gallery and non-gallery programs including an auditorium, bookstore, restaurant, library, education center, and administrative wing. The individual sets of elevators and cores are distributed to guarantee easy access and easy divisibility between zones regulated by different schedules and rules of access.

太原美术馆作为一个建筑群由室内和室外连续和间断的长廊连接。建筑亦作为城市花园景观，鼓励参观者穿过建筑物而不进入博物馆本身。户外斜坡斜插过建筑物，连接各种各样的基础设施、草坪和雕塑花园。从毗邻的汾河的巨大到博物馆的独特空间片断的亲密，建筑与景观的融合表达了不同尺度的地域范围。

博物馆内部空间的安全性通过一个在画廊和包括礼堂、书店、餐厅、图书馆、教育中心、行政翼楼的非画廊之间的高度控制界面来保持。单独和成组的电梯分布设置，以保证各区域之间通行的便利性。

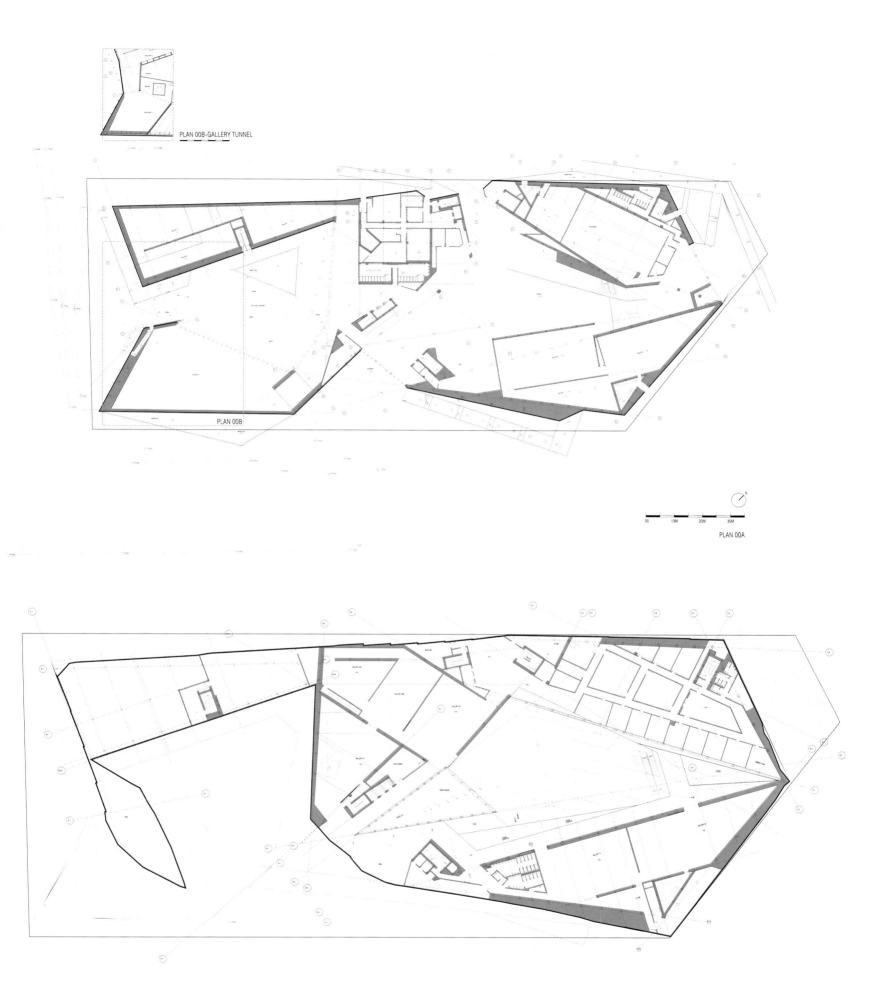

PLAN 00B-GALLERY TUNNEL

PLAN 00B

PLAN 00A

PLAN 03

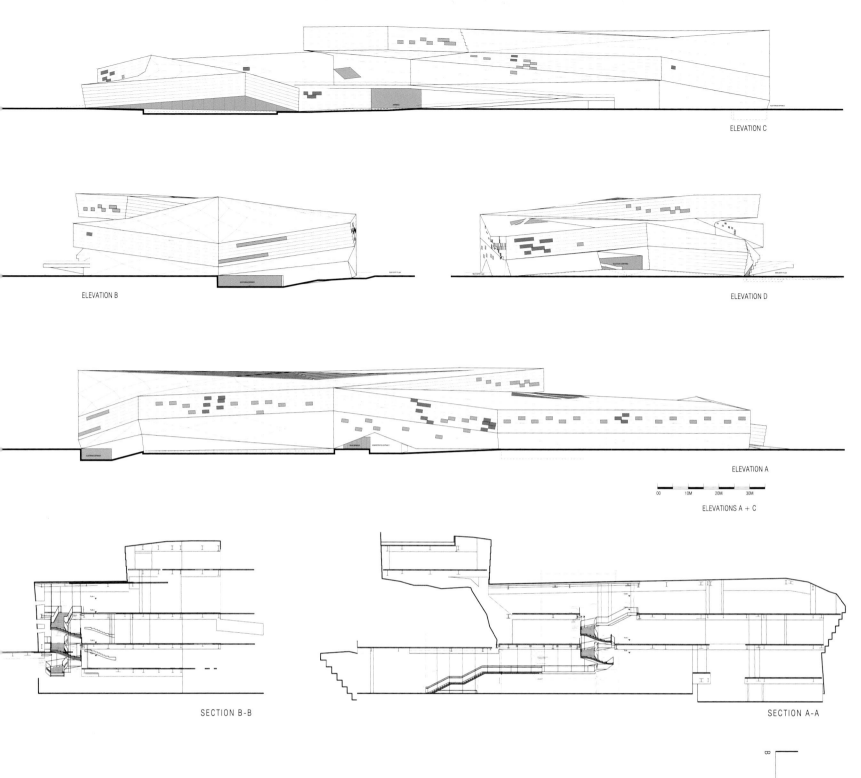

ELEVATION C

ELEVATION B

ELEVATION D

ELEVATION A

ELEVATIONS A + C

00   10M   20M   30M

SECTION B-B

SECTION A-A

The museum galleries are organized to ensure maximum curatorial flexibility. The galleries can be organized into a single, spiraling sequence for large chronological exhibitions or into autonomous clusters operating independently. For visitors architectural cues offer – the placement of ramps and portals, the expansion and contraction of space – provide a means of wayfinding. The building gives visitors the freedom either to follow a predetermined chronological sequence or to skip from one set of galleries to another, in a nonlinear fashion.

Exterior light weight honeycomb panels with stone veneer produce an evocative and elusive material effect and the perception of an exceptional scale. The panels are reflective as if metallic, seemingly too large to be stone panels, but clearly possessing the properties of both materials. Advanced parametric software allowed panels to conform to standard widths, reducing material waste.

博物馆画廊被有系统地组织以确保最大的策展灵活性。对于大型的按时间先后顺序的展览，画廊可以组织成单一的螺旋序列，或独立运作的群。对参观者来说，建筑由坡道和入口的位置，空间的膨胀和收缩提供导视指引。参观者有即可根据预定的时间顺序参观，或以非线性方式从一个画廊跳到另一个画廊参观的自由。

外部轻型蜂窝板和石材面板在引起回忆的同时，产生了难以捉摸的材料效果和独特的尺度感。石板面板如金属板那样反射，看起来太大以至不像是石板，但显然它拥有两种材料的特性。先进的参数软件使得石板符合标准宽度，减少了材料的浪费。

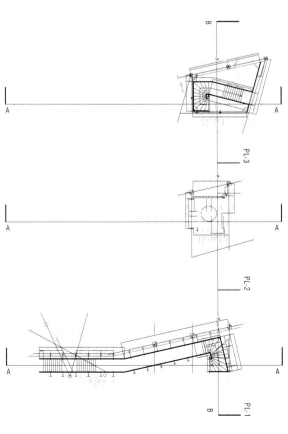

无锡大剧院

Wuxi Grand Theatre

Finnish design is internationally known for its concepts of clear and functional style, which meets the needs of the users from a long term perspective. The first large scale example of Finnish design in China was completed by the opening of the Wuxi Grand Theatre in April 2012. Its design was not only a question of architecture, but the whole design – from functional program to interior design, landscape design, theatre technology, lighting and acoustic design – was controlled by the PES-Architects' team.

The eight gigantic roof wings protect the building for the heat of the sun and create a landmark building with the five facades. The architectural concept is unique: inside the steel wings covered by perforated aluminum panels, there are thousands of LED lights, which make it possible to change the color of the wings according to the character of the performances.

　　芬（兰）式设计在国际上以清晰与实用的设计风格著称，能够从长远的角度满足使用者的需求。2012 年 4 月无锡大剧院的落成标志着中国第一个大规模的芬式设计案例的完工。大剧院的设计不仅仅是建筑设计，而是包括从功能设计到室内设计、景观设计、剧院技术、照明和音响设计在内的整套设计，这些都由 PES 建筑事务所设计团队负责完成。

　　八片大尺度的屋顶翅膀为该建筑物提供遮荫，并创建一个标志性的五立面建筑。其建筑概念是独一无二的：覆盖的穿孔铝板钢铁翼内部有成千个 LED 灯，它们会使"翅膀"根据情景需要改变颜色。

Area · 78,000 m² **Main Materials** · Bamboo, wood **Photography** · Jussi Tiainen, Kari Palsila, Martin Lukasczyk, Pan Weijun

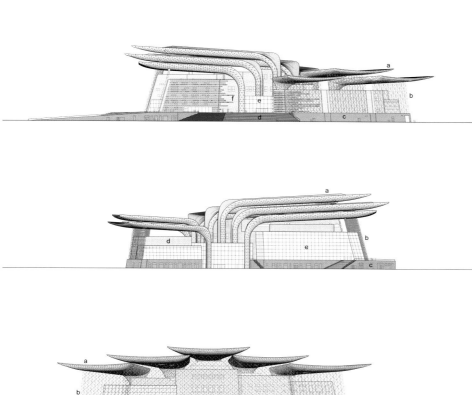

a. wing roofs
b. space truss steel wall
c. building plinth
d. main entrance steps
e. main entrance
f. stone building facade

1:750
South elevation

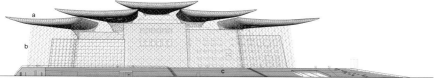

a. wing roofs
b. space truss steel wall
c. building plinth
d. stone building facade
e. glass facade

1:1500
North elevation

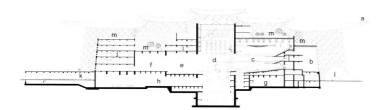

a. wing roofs
b. space truss steel wall
c. building plinth

1:1500
West elevation

a. wing roofs
b. public foyer
c. main auditorium
d. stage tower
e. backstage
f. assembly
g. orchestra lounge
h. assembly and loading
i. rehearsal rooms
j. office / dressing rooms
k. parking hall
l. outdoor theater
m. public roof terraces

1:1500
Section

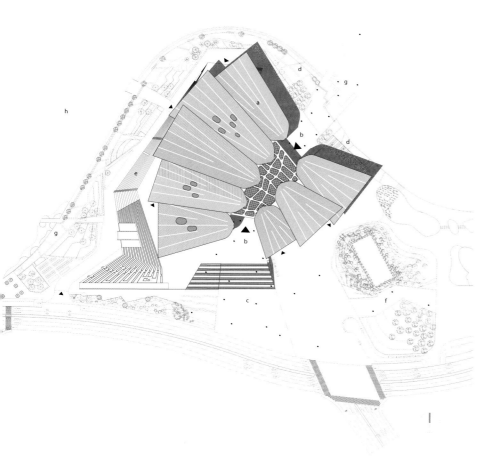

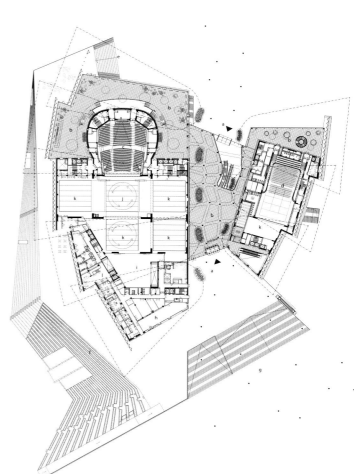

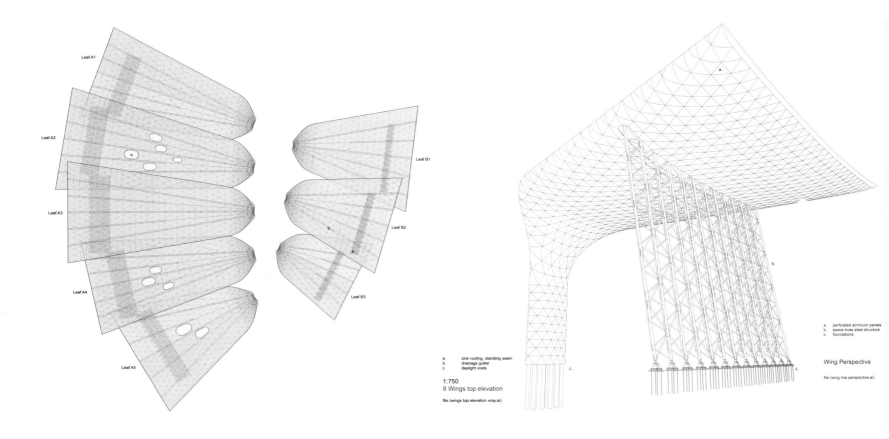

Leaf A1
Leaf A2
Leaf A3
Leaf A4
Leaf A5

Leaf B1
Leaf B2
Leaf B3

a.   zink roofing, standing seam
b.   drainage gutter
c.   daylight voids

1:750
8 Wings top elevation

file (wings top elevation xray.ai)

a
b
c

a.   perforated alminium panels
b.   space truss steel structure
c.   foundations

Wing Perspective

file (wing line perspective.ai)

Another special feature is the "forest" of 50 light columns, each 9 meter high, which start from the main entrance square, support the roof of the central lobby and continue outside of the lakeside entrance into the lake.

There is a strong Chinese feature that runs throughout the whole building: the large scale use of bamboo which is a traditional and sustainable Chinese material. Recently new methods for the production and use of bamboo have made it possible to cover the main opera auditorium with over 17,000 solid bamboo blocks, all individually shaped according to acoustical needs and architectural image. Also for the floor and chairs of main auditorium and in all interior design, the bamboo was used instead of wood.

There is also a material with a Finnish character: almost 20,000 specially designed glass bricks cover the curved wall of the opera auditorium in the lakeside lobby. Finnish nature, lakes and ice, were the architects inspiration.

The Scandinavian interior design tradition is visible in the public lobby areas too, where the furniture and ceiling chandeliers are made of white Corian composite material.

另一个特色是50条9m高的灯柱组成的"森林",它始于主入口广场,支撑着中央大厅的屋顶向外延续到湖畔入口的外面进入湖中。

传统可持续发展的中式材料——竹子的大量使用,使得浓郁的中国特色贯穿整个建筑。生产和使用竹子的最新方法使得以1.7万块实心竹块覆盖主剧院音乐厅成为可能,所有的竹块都是根据声学需要和建筑形象设计成独特的形状。主礼堂的地板和椅子以及所有的室内设计也是用竹材而不是木材。

还有一种具有芬兰特色的材料:将近2万块特制的玻璃砖覆盖在位于湖边大厅音乐厅的弧形墙上。建筑师的灵感来自芬兰的大自然、湖泊与冰。

公共大堂区域的室内设计也有明显的北欧风格,其中的家具和吊灯是由白色可丽耐复合材料制成的。

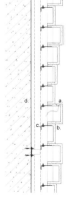

a. integrated LED lighting
b. GRG black wall block
c. secondary structure
d. concrete structure

1:20
Small Auditorium black wall
plan/elevation/section detail

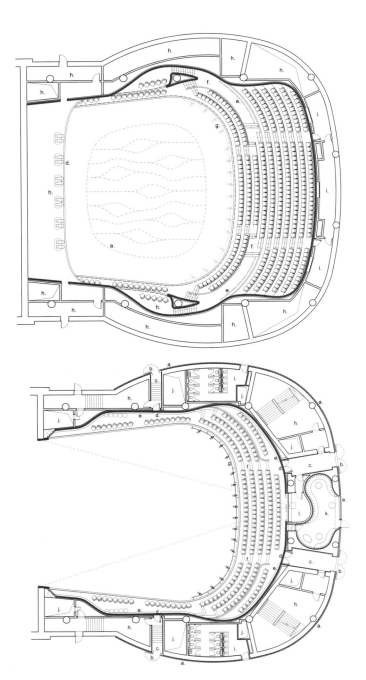

a. golden "opera mask" ceiling reflector
b. procenium lighting bridge
c. royal balcony
d. stage lighting spotlights
e. floor recessed light stripes
f. solid bamboo balcony floor
g. stage lighting bar
h. technical spaces
i. follow spot rooms

1:250
Main Auditorium plan 2nd balcony

a. glass wave wall
b. entrance doors, stainless steel
c. sound buffer frontrooms
d. solid bamboo auditorium doors
e. floor recessed light stripes
f. solid bamboo balcony floor
g. stage lighting bar
h. emercency escape spaces
i. toilets
j. technical spaces
k. VIP room
l. projector room

1:250
Main Auditorium plan 1st balcony

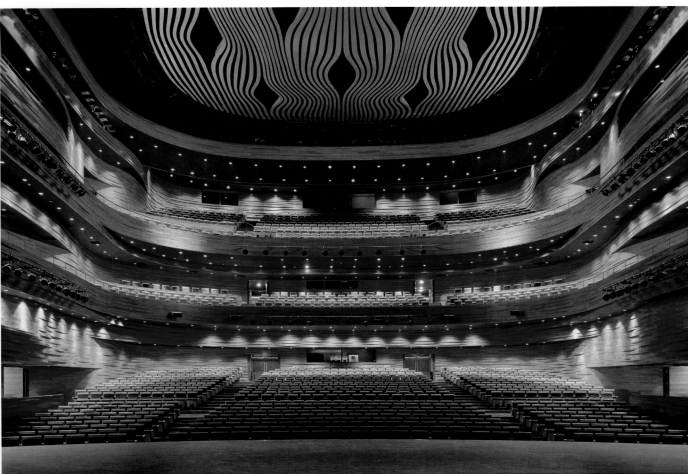

WISA 木质设计旅馆

WISA Wooden Design Hotel

The location, on a skerry washed by the sea, gave rise to the idea of a floating block of wood dashed against the rock in a storm. The hotel has two rectangular-shaped rooms with a freeform exterior space between them. The exterior space is formed by the angle between the rooms and the difference in levels. In theory, the entire building could be made rectangular by straightening out the end walls of the rooms and bringing them up to the same level. The maritime location provided the inspiration for experimenting with boat-building techniques for bending the wooden planks. The curved parts were made by fixing three 8mm layers of pine board on top of each other. The inner and outer layers are unbroken, while the middle layer allows bending. The interior finishes are in birch ply, the structure is in spruce and the elevations in top-grade pine, free from knots. The exterior is treated with a grey protective finish and the interiors are oiled. The building interacts with its surroundings, sometimes standing out from them and sometimes dissolving into them. A piece of timber with a beautiful grey patina is split into laths which bring out the light-colored, organic structure of the wood. What you can see on the rock is a block of wood washed up by the sea.

　　酒店处于一个被海水冲刷的小岛上的位置，给出了在暴风雨中浮动木块撞击岩石的设计理念。酒店有两个矩形的房间以及它们之间的自由外部空间。外部空间由房间和不同层之间夹角形成。理论上，通过使房间的壁变直并把它们砌至相同的水平线上，整个建筑可以建成矩形的。但其海上位置提供了试验弯曲木板的造船技术灵感。弯曲部分是由三块固定的 8mm 的松木板组成，内外层是连续的，而中间层可以弯曲。室内装饰层是桦树板，结构是云杉，正面外观是高品质的无节松树板。外部用灰色保护漆处理，内部用涂彩处理。建筑与周围环境相互作用，有时远离它们有时融入到它们之中。这块覆盖着漂亮灰色铜绿的木板被分成板条状，呈现浅色、有机的木板结构，仿佛在岩石上可以看到的是一块被海水冲上来的木头。

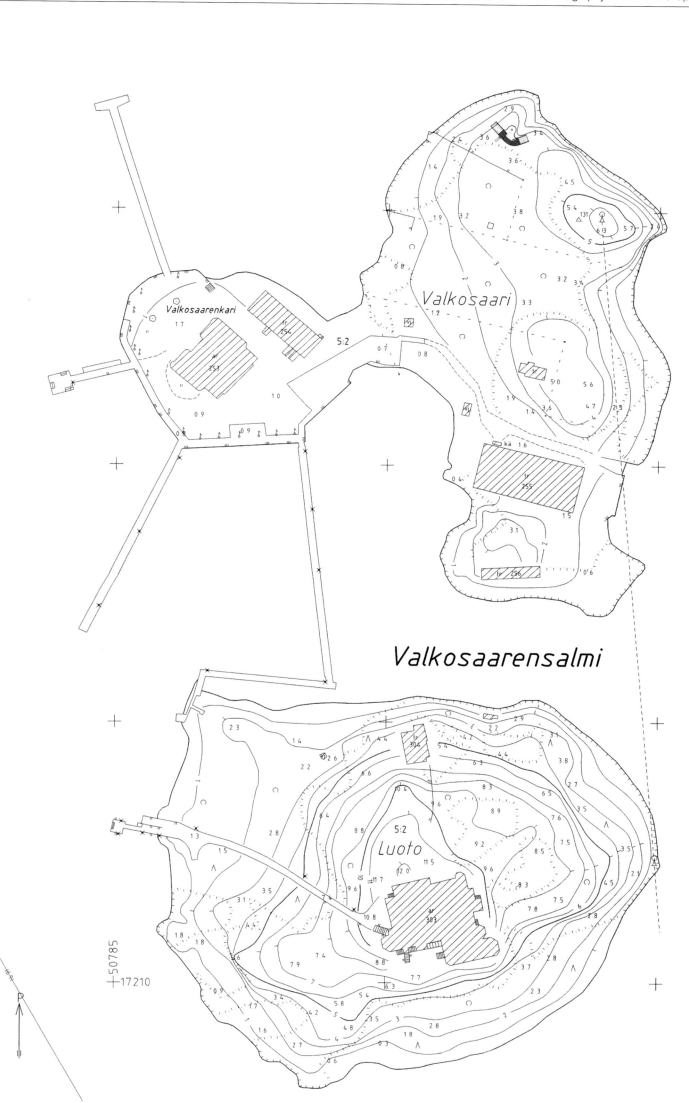

Valkosaari

Valkosaarensalmi

Luoto

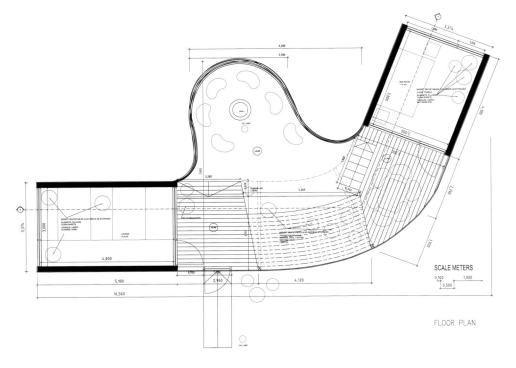

FLOOR PLAN

SCALE METERS

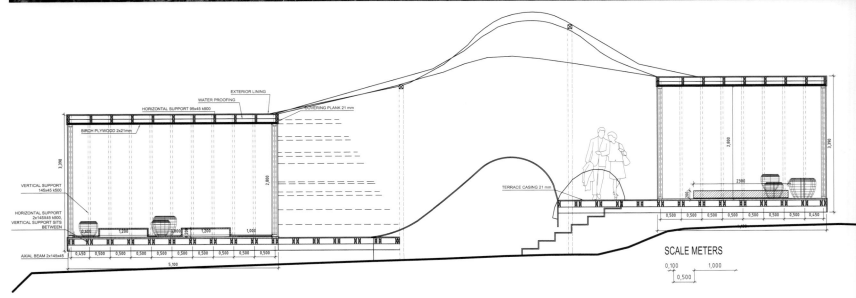

SECTION A-A

SCALE METERS

风之酒店 / 巴塔哥尼亚酒店

Hotel of the Wind / Tierra Patagonia Hotel

The hotel is located at the entrance of "Torres del Paine" National Park, on the shore of Lake Sarmiento.

The water acts as a supporting plane for the splendid Paine massif. Its magnificence makes one think of an extended project in dialogue with the vast territory.

The form seeks to merge with the metaphysical landscape, not to interrupt it. The shape of the hotel is reminiscent of an old fossil, a prehistoric animal beached on the shore of the lake, not unlike those found and studied by Charles Darwin. The building is as though born out of the land, like a fold in the terrain that the wind has carved in the sand. It is anchored to the ground with stone slopes and is completely covered in washed Lenga wood paneling. This finish gives the hotel a silvery sheen. The spatial solution aims for warm and cozy spaces structured by internal pathways, allowing the building to inhabit its extension.

该酒店位于萨米恩托湖岸边的"托雷斯德潘恩"国家公园入口处。

湖水作为壮丽的潘恩地块的支撑平面，该项目的设计灵感源自水面的壮观景色。

酒店形状寻求和无形的景观融为一体，而不是遮挡景观。酒店的形状像一块搁浅在湖边的古老的史前动物化石，跟达尔文发现和研究的那些没什么不同。建筑如同破土而出，像一个在沙地上被风雕刻而成的山坳，固定在石头堆的地面并被水洗智利樱桃木镶板完全覆盖。木镶板的颜色让酒店拥有银色的光泽。空间问题的解决目标是通过内部过道构成温暖舒适的空间，让建筑得以扩展。

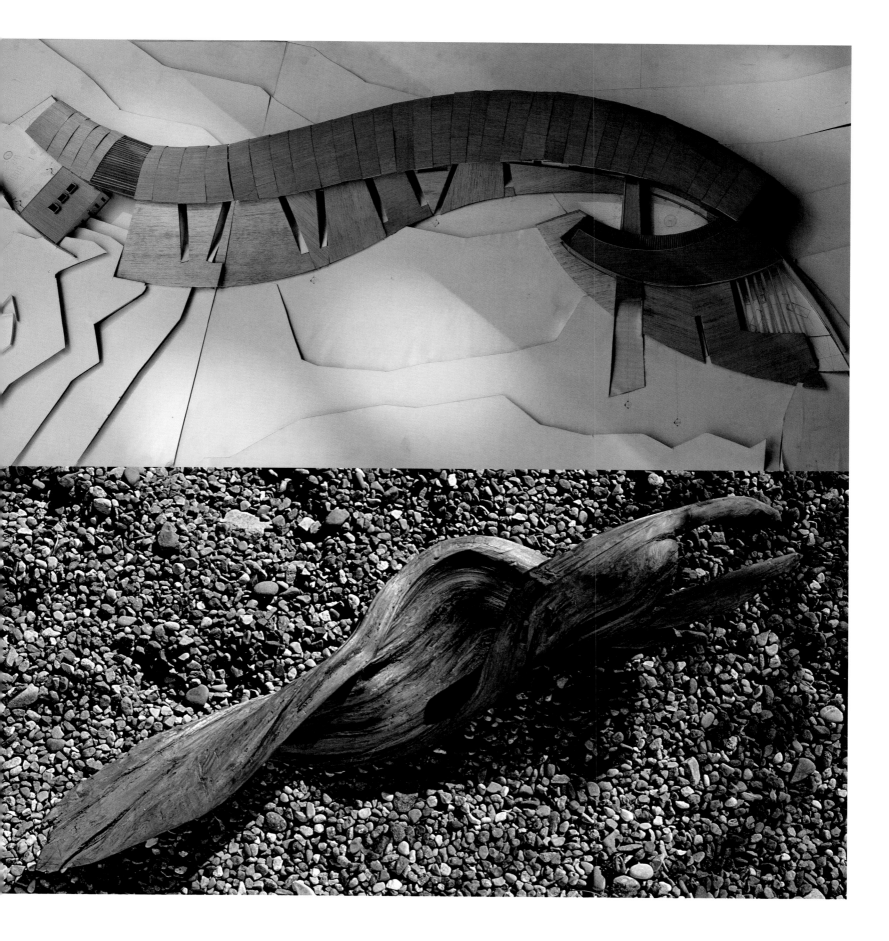

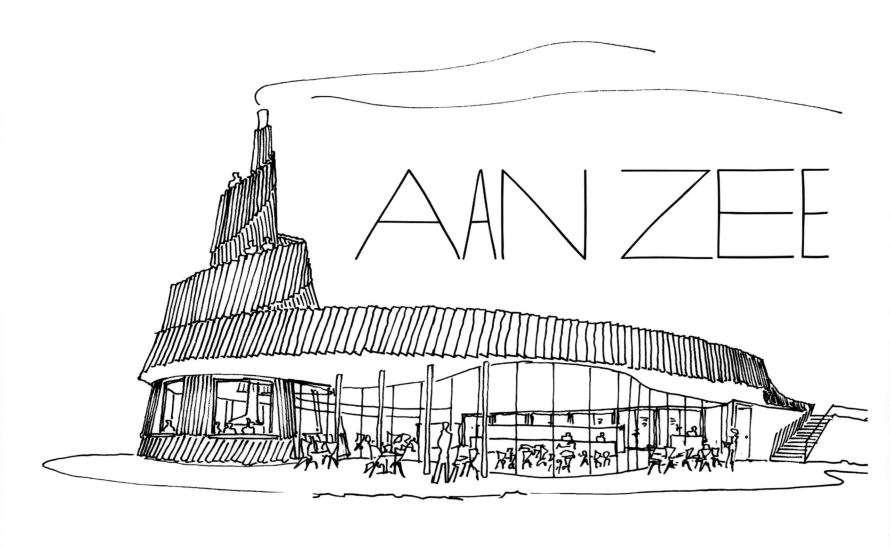

安泽餐厅

Aan Zee

In a dune nature reserve rises a new restaurant: "Aan Zee". It is situated on a breathtaking, rather surreal location: wedged between the dune nature reserve "Voornes duin" and the Rotterdam harbour extension "Maasvlakte", which fills the horizons with heavy industry and rare birds simultaneously. "Aan Zee" is a new restaurant in many ways: It is autarkic in its energy supply, water supply and waste water treatment; It uses sun, wind, wood fire, natural convection and the earth itself as sources for energy, heating, cooling, ventilation, and waste water management. The cook works with a wood burn stove only, and uses ingredients from local farmers and fishermen. The building itself can be fully dismantled and is re-usable both as a building as on a level of materials.

A wooden wall curves up from the dunes, it curls up to form a watchtower. The wood is layered and fixed in overlapping small elements on the facade and roofline. The untreated local wood has been cooked to withstand the elements without chemical treatments, varnish or paint. It will change color over the years, and differently in each direction. The wind and sun will color this building, making it a micro climate report in the sand. This wooden wall encloses the restaurant space, together with an all glass curved facade, aimed at the astonishing views towards the sea. The back of house is made up from concrete standard issue basements, buried under the sand.

在一个沙丘自然保护区出现了一家新餐厅："安泽餐厅"。它位于一个壮观的，相当梦幻般的位置，即沙丘自然保护区"福尔讷沙丘"和鹿特丹港扩建的"Maasvlakte"港之间，重工业和稀有鸟类同时共存的地区。"安泽餐厅"在很多方面都是一个新式餐厅：它在能源供应、供水和废水处理方面是自给自足的；它利用太阳、风、木火、自然对流和地球自身的能源，加热、冷却、通风和进行废水管理。厨师只用一个木火炉做菜，菜单配料来自当地的农民和渔民。建筑本身可以完全被拆除，并作为一个建筑或一定标准的材料重新使用。

木墙从沙丘开始向上弯曲，卷曲起来形成一个瞭望塔。分层的木材重叠固定在外墙和屋顶上面。未经处理的木材，一直要被煮到不用经过化学处理、清漆或涂料就能耐用为止。多年后它的颜色将产生变化，不同方向有不同的变色。风和太阳将为这个建筑上色，使它成为沙地的微气候预报站。木墙围起餐厅空间，连同所有的玻璃弯曲外墙，旨在指向壮丽的海景。房子后面是混凝土标准的地下室，被埋在沙子下面。

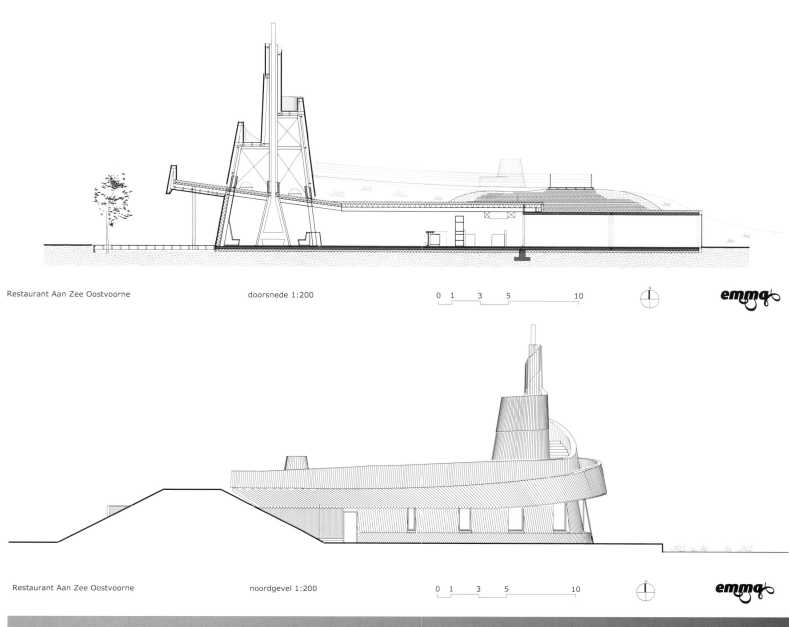

Restaurant Aan Zee Oostvoorne   doorsnede 1:200   0 1 3 5 10   emma

Restaurant Aan Zee Oostvoorne   noordgevel 1:200   0 1 3 5 10   emma

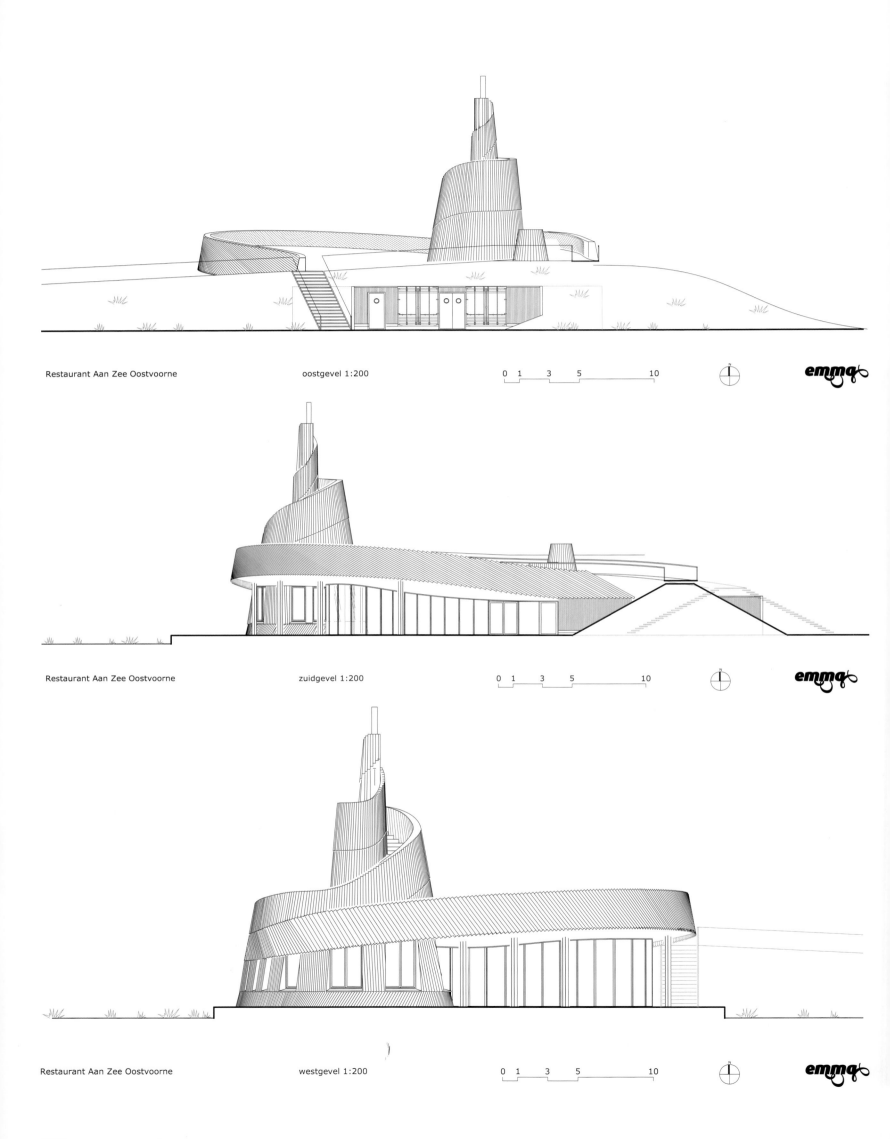

Restaurant Aan Zee Oostvoorne   oostgevel 1:200   0 1   3   5   10   emma

Restaurant Aan Zee Oostvoorne   zuidgevel 1:200   0 1   3   5   10   emma

Restaurant Aan Zee Oostvoorne   westgevel 1:200   0 1   3   5   10   emma

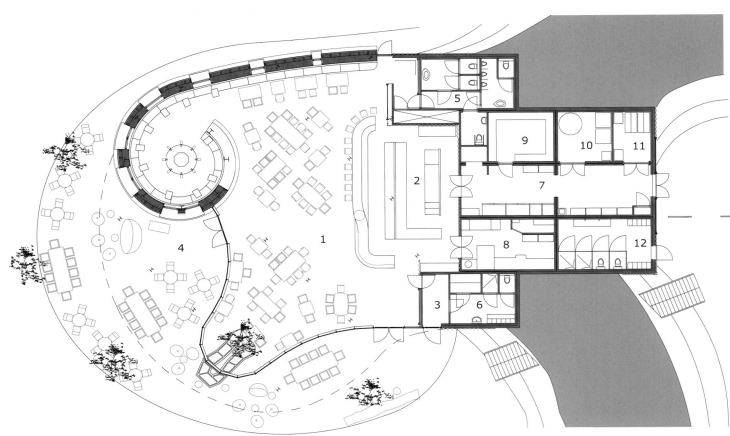

1. restaurant
2. keuken
3. entree
4. terras
5. toiletten
6. personeelsruimte
7. opslag
8. spoelkeuken
9. opslag
10. installatieruimte
11. afvalruimte
12. buitensport faciliteiten

Restaurant Aan Zee Oostvoorne   Begane grond 1:200   0 1 3 5 10

Chalachol 美发沙龙

Chalachol Hair Salon

This salon appears to be wrapped by numerous bamboo sticks hanging from the ceiling in different levels throughout the space. Klinsuwan designed the space as a sculpture that people can walk through and look around, making them question what they are seeing. Instead of creating walls or partitions to separate working areas, the designer divides them by using different lengths of bamboo sticks hanging down from the ceiling until some poles are long enough to hit the floor in order to blend the line between the wall and the ceiling. This also creates permeable walls to screen off the coloring and shampooing areas.

The designer was inspired in this idea from the natural space of caves while he was

traveling in the South of Thailand. In caves, rock formations called stalactites hang downward like icicles, stalagmites are ones that begin below and stick upward. Often, the stalactite and stalagmite will connect, and become a column, forming a wall and creating rooms within the natural space.

Klinsuwan also values the use of materials in this project by using local materials that can control the budget and create an eco friendly place. He focuses on bamboo, a fast-growing material and alternative to timber that is widely used in the region. Klinsuwan collected 11,250 bamboo poles in different sizes from Eastern Thailand, this approach also providing prosperity to local people.

　　该沙龙以这样的形式出现，被无数悬挂在天花板上的竹条包围，竹条以不同的层次遍布于整个空间。科林苏万把空间设计成一个雕塑，人们可以步行通过，四下观望，让他们好奇地发问他们看到的是什么。代替创建墙壁或分成独立的工作区，设计师用不同长度的竹条，从天花板上垂下来直到一些竹条长到接近地面，将墙壁和天花板之间的界限混合起来。这也形成了透风墙壁，隔开了染色区和洗发区。

　　设计师的灵感源自他在泰国南部旅游时看到的洞穴的自然空间。在山洞里，

岩石形成的钟乳石像冰柱似的向下倒挂；石笋是从下往上的。通常，钟乳石和石笋会连起来，成为一个圆柱，在自然空间里形成了墙壁和房间。

　　在本案中，科林苏万也重视材料的使用，通过就地取材，控制预算，创造一个绿色环保的空间。他专注于竹子，这是一种快速生长的、可替代木材的材料，在当地被广泛使用。科林苏万从泰国东部收集了 11250 根大小不同的竹子，这种方法也为当地人们带来了繁荣财富。

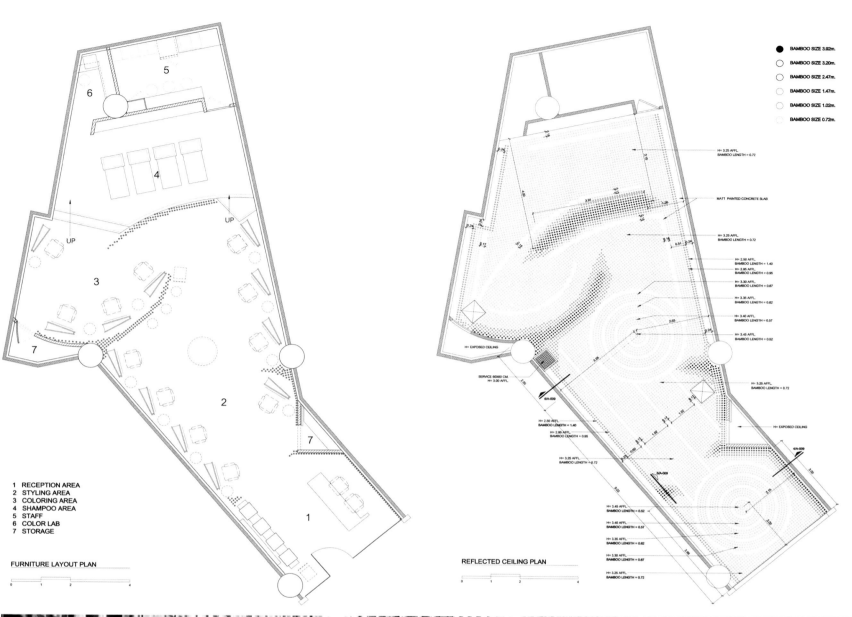

1 RECEPTION AREA
2 STYLING AREA
3 COLORING AREA
4 SHAMPOO AREA
5 STAFF
6 COLOR LAB
7 STORAGE

FURNITURE LAYOUT PLAN

0   1   2   4

● BAMBOO SIZE 3.92m.
○ BAMBOO SIZE 3.20m.
○ BAMBOO SIZE 2.47m.
○ BAMBOO SIZE 1.47m.
○ BAMBOO SIZE 1.02m.
○ BAMBOO SIZE 0.72m.

REFLECTED CEILING PLAN

0   1   2   4

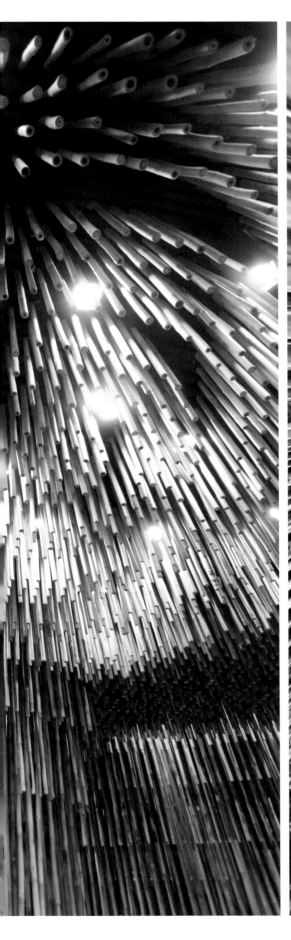

Architect · Perkins + Will   Chief Architect · Jim Huffman   Area · 17,000 m²

VanDusen 植物园游客中心
VanDusen Botanical Garden's Visitor Centre

In 2007, Perkins+Will was commissioned to create a signature, green facility that would increase the VanDusen Botanical Garden's visitorship and enhance its international stature. The Visitor Centre creates a harmonious balance between architecture and landscape – from both a visual and an ecological perspective.

Inspired by the organic forms and natural systems of a native orchid, the project is organized into undulating green roof "petals" that float above rammed earth and concrete walls. These petals and stems are connected by a vegetated land ramp that links the roof to the ground plane, encouraging use of local flora. The building houses a cafe, an expanded library, volunteer facilities, a garden shop, office space and flexible classroom spaces for meetings, lectures, workshops and private functions. Designed to exceed LEED Platinum, the Visitor Centre is pursuing the Living Building Challenge – the most stringent measurement of sustainability in the

built environment. The Visitor Centre uses on-site, renewable sources – geothermal boreholes, solar photovoltaics, solar hot water tubes – to achieve net-zero energy on an annual basis.

Wood is the primary building material, sequestering enough carbon to achieve carbon neutrality. Rainwater is filtered and used for the building's grey water requirements; 100% of black water is treated by an on-site bioreactor and released into a new feature percolation field and garden. Natural ventilation is assisted by a solar chimney, composed of an operable glazed oculus and an aluminum heat sink, which converts the sun's rays to convection energy. Located in the centre of the atrium, and exactly at the centre of all the building's various radiating geometry, the solar chimney highlights the role of sustainability by form and function.

在 2007 年，Perkins + Will 受委托设计一个显著的绿色设施，以增加 VanDuen 植物园的游客数量和提高其国际地位。此游客中心同时从视觉和生态视角上创造了建筑与景观之间的和谐平衡。

灵感来自原生兰花的动感和自然形式，该案被组织成起伏的绿色屋顶"花瓣"，浮在夯土和混凝土墙上。这些花瓣和茎通过连接屋顶与地面的植物斜坡连在一起，促进当地植物群的应用。这栋房子有咖啡馆、扩建的图书馆、志愿者设施、花园商店、办公空间以及用于会议、讲座、作坊和私用功能的灵活的课堂空间。为了超越 LEED 白金资格，游客中心正追求生态建筑挑战——可持续发展的最严格的认证计划。游客

中心使用就地取材的可再生能源——地热井、太阳能光伏电池、太阳能热水管，实现年度净零能耗。

木材作为主要的建筑材料，分离了足够的碳排放以实现碳中和。雨水被过滤以满足建筑的灰水需求；100% 黑水经生物反应器处理并释放到一个新型的过滤场和花园。自然通风装置由太阳能烟囱协助，烟囱由一个可操作的琉璃圆窗和铝散热片组成，将太阳光线转换成对流能源。太阳能烟囱位于正厅的中央，也恰好在所有建筑物的各种辐射状的几何中心，它在形式和功能上都凸显了可持续性发展的作用。

PERKINS
+WILL

VanDusen Botanical Garden Visitor Centre
Vancouver Board of Parks and Recreation
Vancouver, BC

Only MATERIALS that efficiently utilize resources are used in the project.

An operable GLAZED OCULUS with a solar heat sink allows for passive ventilation.

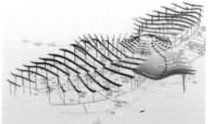

CONTEXT Located at Oak and 37th Street, VanDusen is less than 5 km from downtown Vancouver and is looking to draw visitors from an increased street presence.

PROGRAM The 1,765 SM facility houses a cafe, an expanded library and volunteer facilities, a new garden shop, administration and flexible classroom space.

RHINO SOFTWARE was used to design and 3D model the undulating roof forms.

The complex petal ROOF STRUCTURE utilizes a glulam post-and-beam assembly.

FLOWER AS METAPHOR
Rooted in place and yet...
harvests all its own energy + water
is adapted to climate + site
operates pollution free
promotes health + well-being
comprised of integrated systems
embodies beauty

INNOVATION The VanDusen Botanical Garden Visitor Centre is tailored to its stunning natural context, creating a harmonious balance between built and natural landscapes and demonstrating the positive co-evolution of natural and human systems.

The petal-shaped green roofs contain native planting that encourages use by local fauna, butterfly meadows and rainwater collection areas. An extensive water management system includes rainwater, stormwater and blackwater strategies.

Paviljoen Puur 展厅

Paviljoen Puur

The pavilion is named after its operator: Puur Produkties. He signs for taste: with a passion for surprise and drive for quality this company distinguishes itself for nearly ten years in the greater Amsterdam urban area.

The interior of Paviljoen Puur is sturdy and authentic, as are its surroundings. Owner Sander Walta combined the furniture of the colorful Italian design brand Moroso with traditional wooden elements by Simons Houtwerk and innovative design lighting by Flos. The interior is conceived as extremely flexible, for any type of event, an own customized lay out can be arranged.

The new green is eco-chic: a combination of sustainable and stylish. This is evident from the materials used for the building and its interiors. The use of air heat pumps and biologically purified water emphasizes the sustainable nature of Paviljoen Puur. The new green is a combination of both nature and city.

Paviljoen Puur is located on the exact site of a former soldiers shelter. The footprint of this shelter was taken as the basis for the pavilion. Around its base a wooden wall curves up to protect the site and encompass the program. The design is inspired by the undulating forms of the slopes and the surrounding landscape.

该馆以其经营者 Puur Produkties 命名。他是品位的标志：拥有让人惊讶的热情，带动着公司近十年来在更大的阿姆斯特丹商圈与众不同的特色。

Paviljoen Puur 展厅内部是坚固真实的，如同它周围的环境一样。业主桑德·沃尔塔将色彩丰富的具有西蒙斯·豪特沃克的传统木元素的意大利设计品牌 Moroso 家具以及弗洛斯的创新照明设备结合在一起。内部设计是无比灵活的，适用于任何类型的活动，也可自行安排定制布局。

新绿色是生态别致的：是一种可持续发展和时尚的结合，这一点从建筑用料及内部装饰都很明显的表现出来。空气热泵的使用及生物净化水强调了 Paviljoen Puur 的可持续性本质。新绿色也是大自然和城市结合。

Paviljoen Puur 正好位于前士兵收容所的位置。这个收容所的足迹成为了展馆的基础。在基地周围一面木墙弯曲向上保护着这个遗址并围绕着整个项目。设计的灵感来自于斜坡的起伏形状和周围景观。

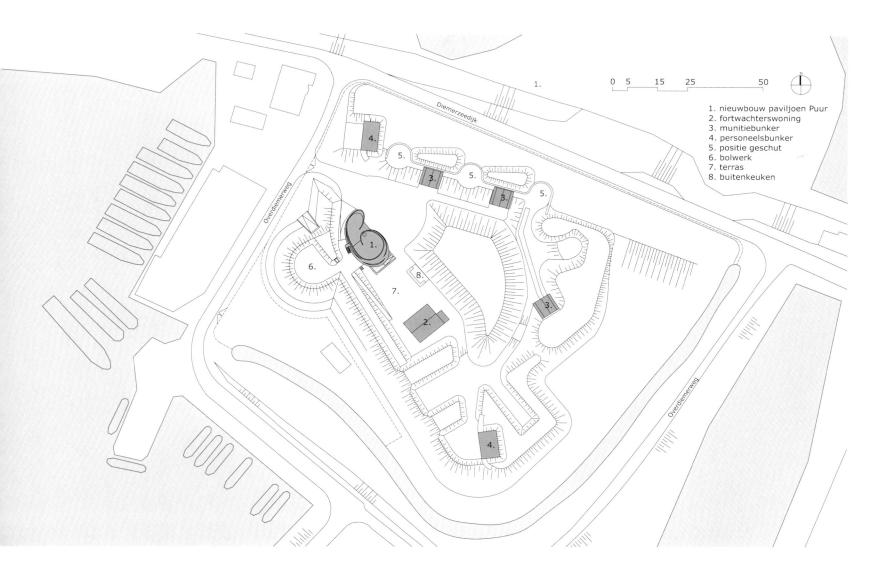

1.

0 5   15   25   50   N

1. nieuwbouw paviljoen Puur
2. fortwachterswoning
3. munitiebunker
4. personeelsbunker
5. positie geschut
6. bolwerk
7. terras
8. buitenkeuken

Fort bij Diemerdam   situatie 1:1000   emma

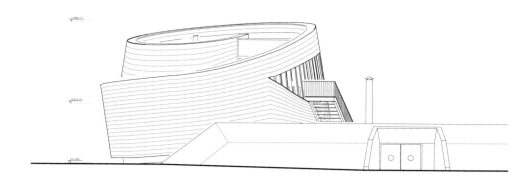

0   1   3   5   10

Fort bij Diemerdam   Noordgevel 1:150

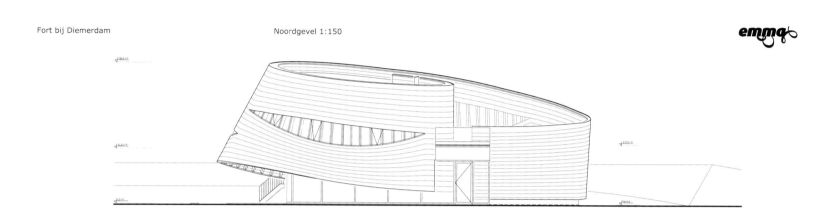

0   1   3   5   10

Fort bij Diemerdam   Oostgevel 1:150

0   1   3   5   10

Fort bij Diemerdam   Westgevel 1:150

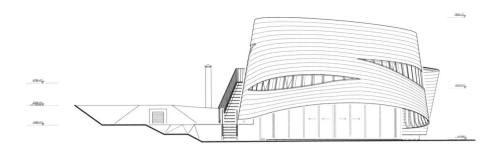

Fort bij Diemerdam   Zuidgevel 1:150   emma

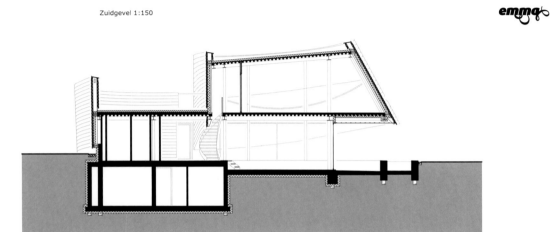

Fort bij Diemerdam   Doorsnede A 1:150   emma

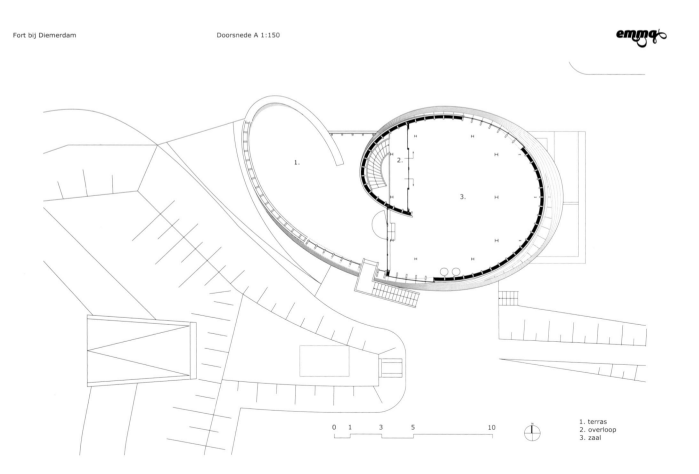

1. terras
2. overloop
3. zaal

Fort bij Diemerdam   Verdieping 1:150   emma

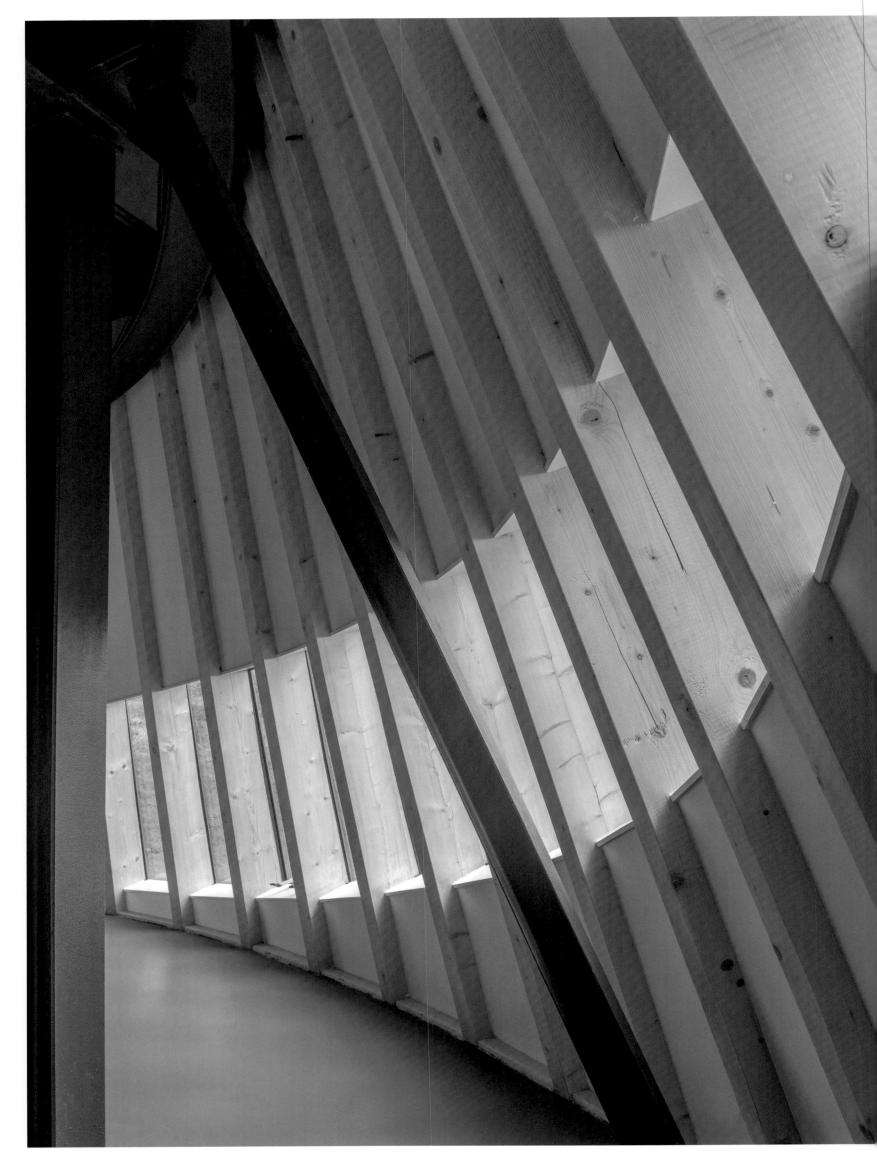

CADE 葡萄酒厂
CADE Winery

CADE Winery comprises two buildings and a barrel storage cave. The main concept for the design of the winery building represents a bold approach to the process of winemaking, a new slant on a well-established practice.

Inspired by the location – the dramatic rolling hills of Howell Mountain – Fernandez incorporated the scenic surroundings into the design. The flow and shape of the retaining walls reflect the silhouette of the hillside, and the vertical proportions of the buildings are complemented by the tall trees that surround them. The cave design – 14,500 square feet built into the hillside – was inspired by the inverted version of the PlumpJack logo.

Green design was a necessity, which features building materials composed of concrete with 30 percent fly ash, steel made up of 98 percent recycled material, glass and wood certified by the Forest Stewardship Council (FSC). The main building at CADE Winery relies on natural ventilation and its own concrete mass for cooling.

CADE 葡萄酒厂由两栋建筑和一个储存仓构成。酒厂建筑设计的主要构想体现了酿酒过程中的一个大胆的、行之有效的做法。

设计灵感来自豪厄尔山的位置，急剧起伏的山丘，费尔南德兹将周围风景纳入设计中，挡土墙的流动和形状反映了山坡的轮廓，建筑的垂直部分和四周的高树相辅相成。储存仓建在山坡上，面积为 1347 m²，灵感来自 Plump Jack 的商标的倒置版。

环保设计是必要的，主要特色体现在建筑材料上：含 30% 飞尘的混凝土；由 98% 回收材料组成的钢材；经森林管理委员会（FSC）认证的玻璃和木材。CADE 酒厂的主楼依靠自然通风和自身的混凝土块降温。

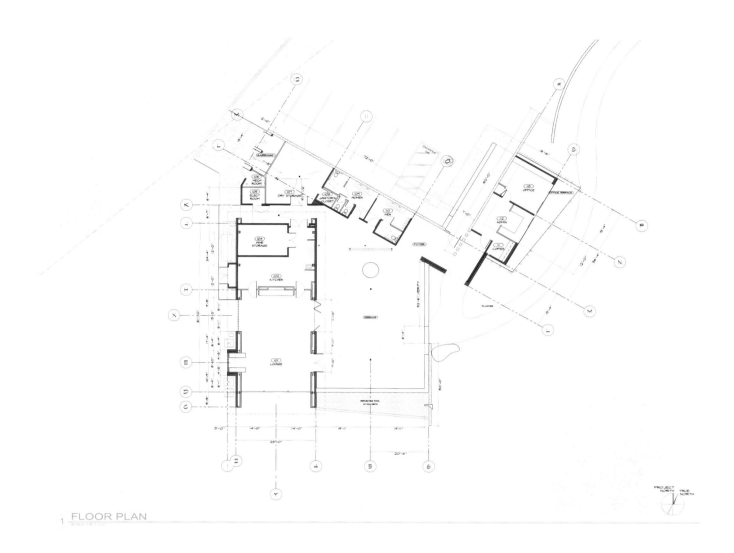

1 FLOOR PLAN

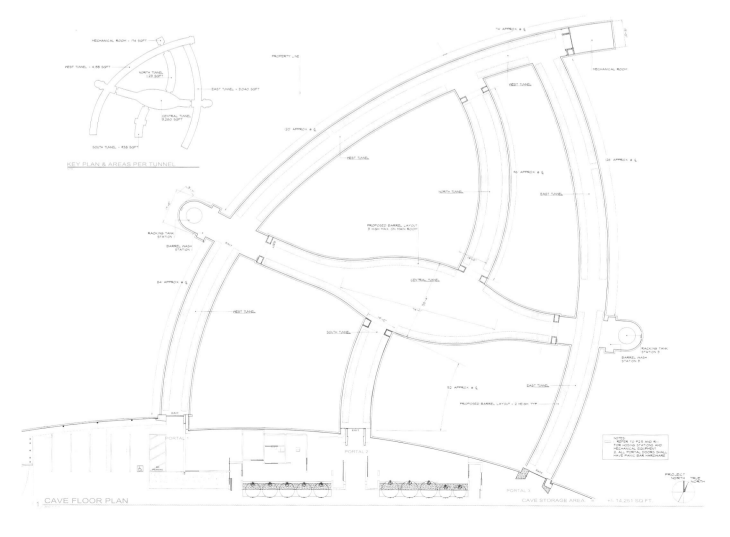

KEY PLAN & AREAS PER TUNNEL

1 CAVE FLOOR PLAN

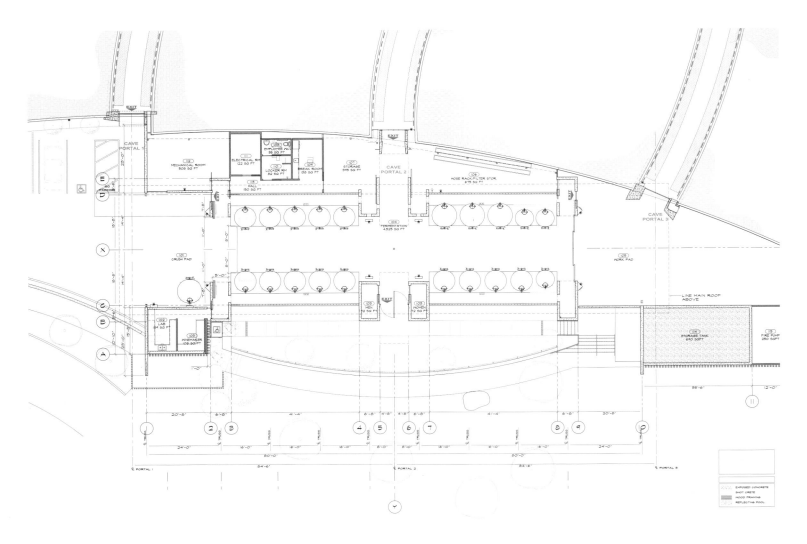

1 FLOOR PLAN
SCALE: 1/8" = 1'-0"

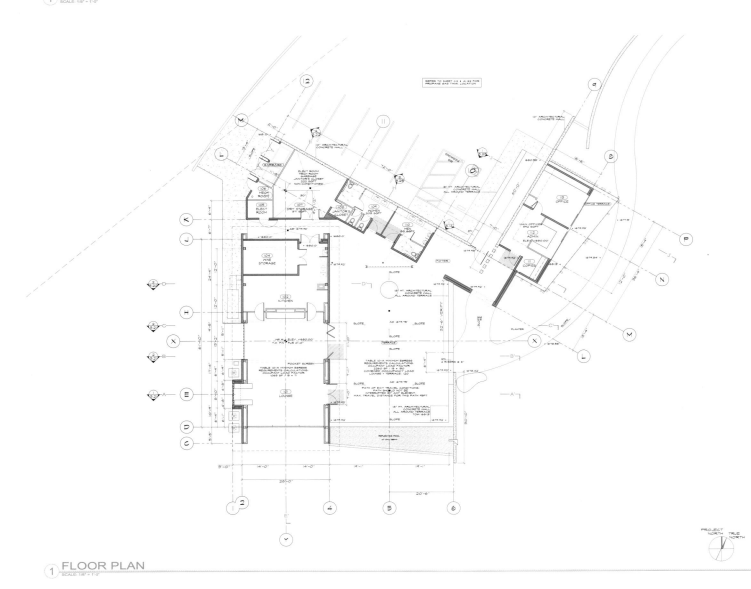

1 FLOOR PLAN
SCALE: 1/8" = 1'-0"

安普里奥店
Emporio

Located next to the meat factory and products of the soil of the Chilean company Coexca SA, the project is thought to be the showcase of the brand deli "Field Noble" like an eye looking towards the main road and the public. The eye symbolized by a glass surface is surrounded by walls, wooden floors and ceilings, creating a flight on the main structure and allows, as a lid, regular amount of light has to pass through the windows. Outside, this flight protects terraces surrounding the building. The control of light and heat is a key to

protect perishable products sold in the store. Also, the front of outdoor wooden air chamber acts as a sunscreen skin.

As in all projects Infiniski Architecture tries to adapt to the conditions of natural environment of the site, for maximum eco-efficiency of the home. With the joint action of passive solar design, cross-ventilated facades and the use of materials insulation quality, high quality building thermal achieved.

　　项目位于肉厂和智利土壤产品公司 Coexca SA 旁边，用于展示熟食品牌 "Field Noble"，它像一只眼睛在看着主道和公共空间。代表眼睛的玻璃被墙壁、木地板和天花板包围，形成了主要结构的一段楼梯，作为一个盖子让连续光线通过窗户进来。在户外，这段楼梯保护了该建筑周围的台阶。光和热的控制是保护商店

出售的易变质产品的关键。同时，户外木质空调房的前面起到了防晒的作用。

　　在所有项目中，Infiniski 建筑事务所都试图适应场地的自然环境条件，尽可能地取得最大的生态效率。与诱导式太阳能设计相结合，交叉通风立面和优质绝缘材料的使用，形成了高效的热能建筑。

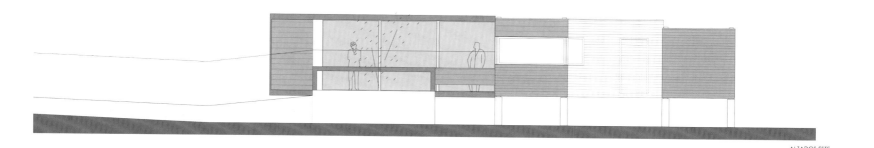

ALZADOS ESTE

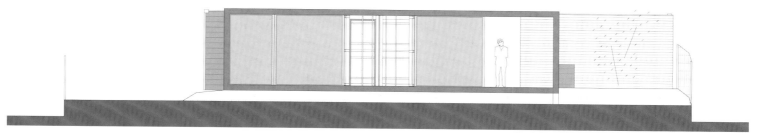

ALZADOS SUR

INFINISKI / James&mau / EMPORIO
ArquiteCtuRa i ConstrucciòN sostenibles   Talca (Chile) _ Alzados _ e.1/100

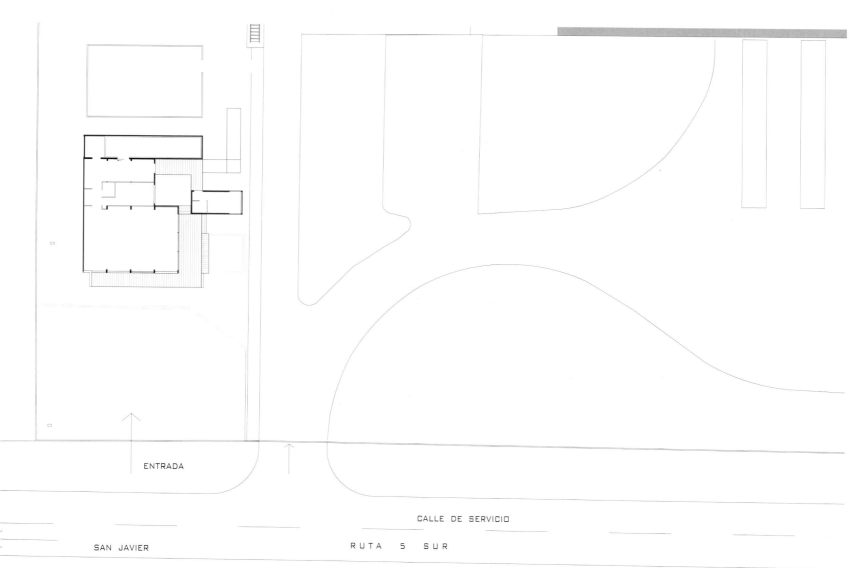

NFINISKI / **James&mau** / EMPORIO
ArQuitEctuRa i CoNstRuccióN sostenibles / arquitectura / Talca (Chile) _ Planta de situación _ e.1/300

ENTRADA

CALLE DE SERVICIO

SAN JAVIER   RUTA 5 SUR

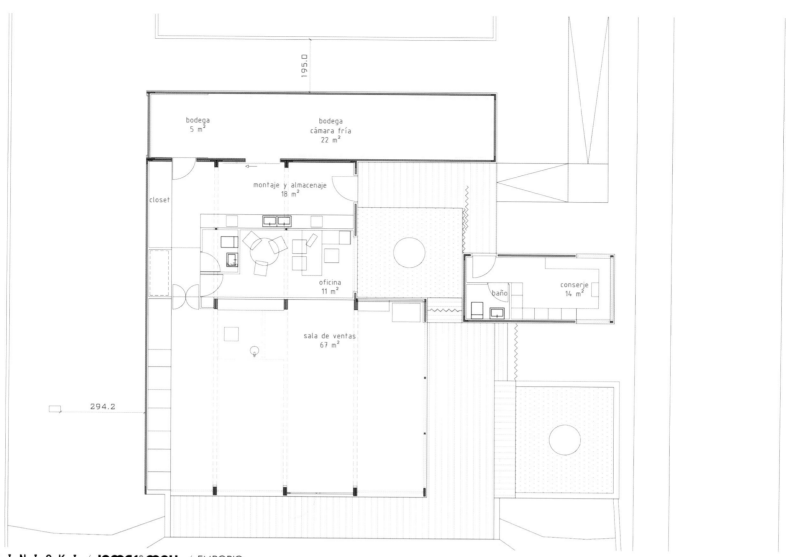

bodega
5 m²

bodega
cámara fría
22 m²

195.0

closet

montaje y almacenaje
18 m²

oficina
11 m²

baño

conserje
14 m²

sala de ventas
67 m²

294.2

INFINISKI / **james&mau** / EMPORIO
ArQuitEctuRe I CoNstRuccIóN sostenibles   arquitectura   Talca (Chile) _ Planta de distribución _ e.1/100

意大利拉奎拉德尔帕克礼堂

Auditorium del Parco, L'Aquila, Italy

Creating an illusion of instability, the auditorium is formed by three interconnected cubes made entirely of wood (1,165 cubic meters in total) that ironically appears as they had "haphazardly tumbled down" and came to rest upon each other. The entire structure was prefabricated and then assembled onsite by Log Engineering, who pieced it together with 800,000 nails, 100,000 screws and 10,000 brackets.

Although the wooden cubes provide a striking contrast to the neighboring, 16th-century Castle of L'Aquila, the material was chosen to optimize the building's acoustic function. Using larch from Val di Fiemme in Trento, which is highly-valued and famously known for being used by Cremona's 17th-century master lute-makers, Stradivarius being the most famous. The building is intended to perform like a musical instrument.

The 2,500 square meter structure features a central volume, which houses the 238-seat auditorium, and two service volumes: a public service area with a foyer and a performance service area with dressing rooms and additional support space. Each is clad in multicolored and specially treated larch tiles, measuring around 25 centimeters wide and 4 centimeters thick. The 16 sides of the cubes that can be seen are not all equal but vary depending on various, alternating architectonic criteria that give the structure a light, lively, and vibrant look. For example, a glass encased staircase punctures through the wooden facade of the foyer as it leads to a second story, while an opaque surface on the south side and roof softens natural light to illuminate a transition space between the auditorium and foyer. In addition, the walls' raw wood surfaces are hung with a series of acoustic panels orientated towards the audience to reflect sound inside the auditorium.

德尔帕克礼堂由三块相互关联的木质方块组成（总体积为 1 165 m³），给人一种不稳的错觉，颇具讽刺意味，它们看起来像"随意地跌下来"，而后一块挨着一块立在地面上。整个建筑物预先制好，再由罗格工程公司通过 800 000 颗钉子、100 000 颗螺丝钉和 10 000 个支架现场组装而成。

尽管这些木质立方体与相邻的 16 世纪的拉奎拉城堡形成了鲜明的对比，选择这些材料主要是为了优化建筑物的声学功能。它使用了来自费耶美山谷的落叶松木，这种木材很宝贵，并因曾被 17 世纪克雷莫纳市最著名的制琴大师特拉迪瓦里使用而闻名于世。这个建筑物便试图以乐器的形状表现自己。

该建筑物面积为 2 500 m²，中央体量有一个设有 238 个座位的礼堂，此外还有两个服务区：一个带门厅的公共服务区域和一个带更衣室及额外支持空间的表演服务区。每个外层都覆以彩色的落叶松质瓦片，它们经过了特殊的处理、大约 25 cm 宽、4 cm 厚。这些建筑体有 16 面可见，但并非每面都相等，而是取决于不同的建筑标准，这使得建筑外观明亮而活泼，充满活力。例如，门厅的木制外立面有玻璃包围的楼梯穿过，直接通往二层，而南面不透明的表面和屋顶柔和了自然光线，照亮了礼堂和门厅之间的过渡空间。此外，礼堂墙壁的原木表面悬挂着一系列的声学面板，它们朝向观众，进行声波的反射。

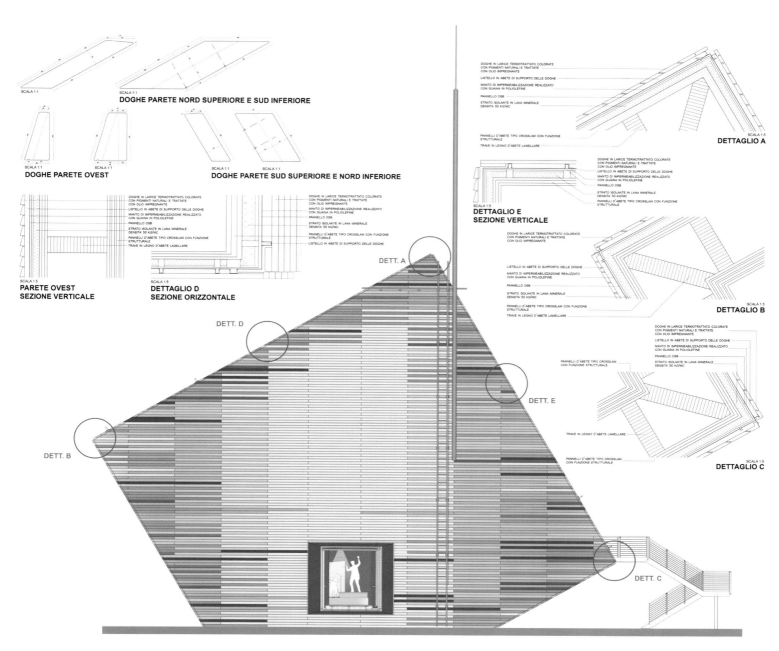

SCALA 1:1

DOGHE PARETE NORD SUPERIORE E SUD INFERIORE

SCALA 1:1   SCALA 1:1

DOGHE PARETE OVEST

SCALA 1:1   SCALA 1:1

DOGHE PARETE SUD SUPERIORE E NORD INFERIORE

DOGHE IN LARICE TERMOTRATTATO COLORATE
CON PIGMENTI NATURALI E TRATTATE
CON OLIO IMPREGNANTE
LISTELLO IN ABETE DI SUPPORTO DELLE DOGHE
MANTO DI IMPERMEABILIZZAZIONE REALIZZATO
CON GUAINA IN POLIOLEFINE
PANNELLO OSB
STRATO ISOLANTE IN LANA MINERALE
DENSITÀ 50 KG/MC
PANNELLI D'ABETE TIPO CROSSLAM CON FUNZIONE
STRUTTURALE
TRAVE IN LEGNO D'ABETE LAMELLARE

DOGHE IN LARICE TERMOTRATTATO COLORATE
CON PIGMENTI NATURALI E TRATTATE
CON OLIO IMPREGNANTE
MANTO DI IMPERMEABILIZZAZIONE REALIZZATO
CON GUAINA IN POLIOLEFINE
PANNELLO OSB
STRATO ISOLANTE IN LANA MINERALE
DENSITÀ 50 KG/MC
PANNELLI D'ABETE TIPO CROSSLAM CON FUNZIONE
STRUTTURALE
LISTELLO IN ABETE DI SUPPORTO DELLE DOGHE

SCALA 1:5
**PARETE OVEST
SEZIONE VERTICALE**

SCALA 1:5
**DETTAGLIO D
SEZIONE ORIZZONTALE**

DOGHE IN LARICE TERMOTRATTATO COLORATE
CON PIGMENTI NATURALI E TRATTATE
CON OLIO IMPREGNANTE
LISTELLO IN ABETE DI SUPPORTO DELLE DOGHE
MANTO DI IMPERMEABILIZZAZIONE REALIZZATO
CON GUAINA IN POLIOLEFINE
PANNELLO OSB
STRATO ISOLANTE IN LANA MINERALE
DENSITÀ 50 KG/MC
PANNELLI D'ABETE TIPO CROSSLAM CON FUNZIONE
STRUTTURALE
TRAVE IN LEGNO D'ABETE LAMELLARE

SCALA 1:5
DETTAGLIO A

DOGHE IN LARICE TERMOTRATTATO COLORATE
CON PIGMENTI NATURALI E TRATTATE
CON OLIO IMPREGNANTE
LISTELLO IN ABETE DI SUPPORTO DELLE DOGHE
MANTO DI IMPERMEABILIZZAZIONE REALIZZATO
CON GUAINA IN POLIOLEFINE
PANNELLO OSB
STRATO ISOLANTE IN LANA MINERALE
DENSITÀ 50 KG/MC
PANNELLI D'ABETE TIPO CROSSLAM CON FUNZIONE
STRUTTURALE

SCALA 1:5
**DETTAGLIO E
SEZIONE VERTICALE**

DOGHE IN LARICE TERMOTRATTATO COLORATE
CON PIGMENTI NATURALI E TRATTATE
CON OLIO IMPREGNANTE
LISTELLO IN ABETE DI SUPPORTO DELLE DOGHE
MANTO DI IMPERMEABILIZZAZIONE REALIZZATO
CON GUAINA IN POLIOLEFINE
PANNELLO OSB
STRATO ISOLANTE IN LANA MINERALE
DENSITÀ 50 KG/MC
PANNELLI D'ABETE TIPO CROSSLAM CON FUNZIONE
STRUTTURALE
TRAVE IN LEGNO D'ABETE LAMELLARE

SCALA 1:5
DETTAGLIO B

DOGHE IN LARICE TERMOTRATTATO COLORATE
CON PIGMENTI NATURALI E TRATTATE
CON OLIO IMPREGNANTE
LISTELLO IN ABETE DI SUPPORTO DELLE DOGHE
MANTO DI IMPERMEABILIZZAZIONE REALIZZATO
CON GUAINA IN POLIOLEFINE
PANNELLO OSB
STRATO ISOLANTE IN LANA MINERALE
DENSITÀ 50 KG/MC

PANNELLI D'ABETE TIPO CROSSLAM
CON FUNZIONE STRUTTURALE

TRAVE IN LEGNO D'ABETE LAMELLARE

PANNELLI D'ABETE TIPO CROSSLAM
CON FUNZIONE STRUTTURALE

SCALA 1:5
DETTAGLIO C

DETT. A

DETT. D

DETT. E

DETT. B

DETT. C

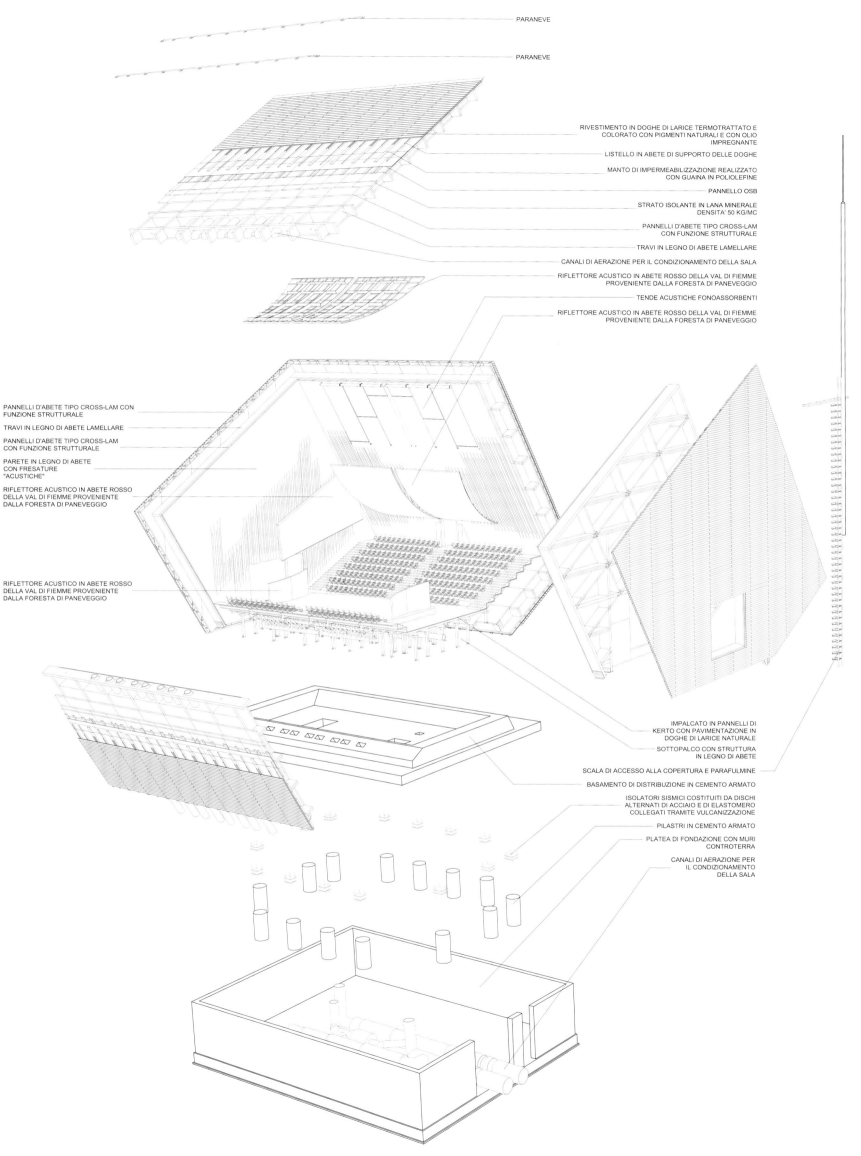

PARANEVE

PARANEVE

RIVESTIMENTO IN DOGHE DI LARICE TERMOTRATTATO E
COLORATO CON PIGMENTI NATURALI E CON OLIO
IMPREGNANTE

LISTELLO IN ABETE DI SUPPORTO DELLE DOGHE

MANTO DI IMPERMEABILIZZAZIONE REALIZZATO
CON GUAINA IN POLIOLEFINE

PANNELLO OSB

STRATO ISOLANTE IN LANA MINERALE
DENSITA' 50 KG/MC

PANNELLI D'ABETE TIPO CROSS-LAM
CON FUNZIONE STRUTTURALE

TRAVI IN LEGNO DI ABETE LAMELLARE

CANALI DI AERAZIONE PER IL CONDIZIONAMENTO DELLA SALA

RIFLETTORE ACUSTICO IN ABETE ROSSO DELLA VAL DI FIEMME
PROVENIENTE DALLA FORESTA DI PANEVEGGIO

TENDE ACUSTICHE FONOASSORBENTI

RIFLETTORE ACUSTICO IN ABETE ROSSO DELLA VAL DI FIEMME
PROVENIENTE DALLA FORESTA DI PANEVEGGIO

PANNELLI D'ABETE TIPO CROSS-LAM CON
FUNZIONE STRUTTURALE

TRAVI IN LEGNO DI ABETE LAMELLARE

PANNELLI D'ABETE TIPO CROSS-LAM
CON FUNZIONE STRUTTURALE

PARETE IN LEGNO DI ABETE
CON FRESATURE
"ACUSTICHE"

RIFLETTORE ACUSTICO IN ABETE ROSSO
DELLA VAL DI FIEMME PROVENIENTE
DALLA FORESTA DI PANEVEGGIO

RIFLETTORE ACUSTICO IN ABETE ROSSO
DELLA VAL DI FIEMME PROVENIENTE
DALLA FORESTA DI PANEVEGGIO

IMPALCATO IN PANNELLI DI
KERTO CON PAVIMENTAZIONE IN
DOGHE DI LARICE NATURALE

SOTTOPALCO CON STRUTTURA
IN LEGNO DI ABETE

SCALA DI ACCESSO ALLA COPERTURA E PARAFULMINE

BASAMENTO DI DISTRIBUZIONE IN CEMENTO ARMATO

ISOLATORI SISMICI COSTITUITI DA DISCHI
ALTERNATI DI ACCIAIO E DI ELASTOMERO
COLLEGATI TRAMITE VULCANIZZAZIONE

PILASTRI IN CEMENTO ARMATO

PLATEA DI FONDAZIONE CON MURI
CONTROTERRA

CANALI DI AERAZIONE PER
IL CONDIZIONAMENTO
DELLA SALA

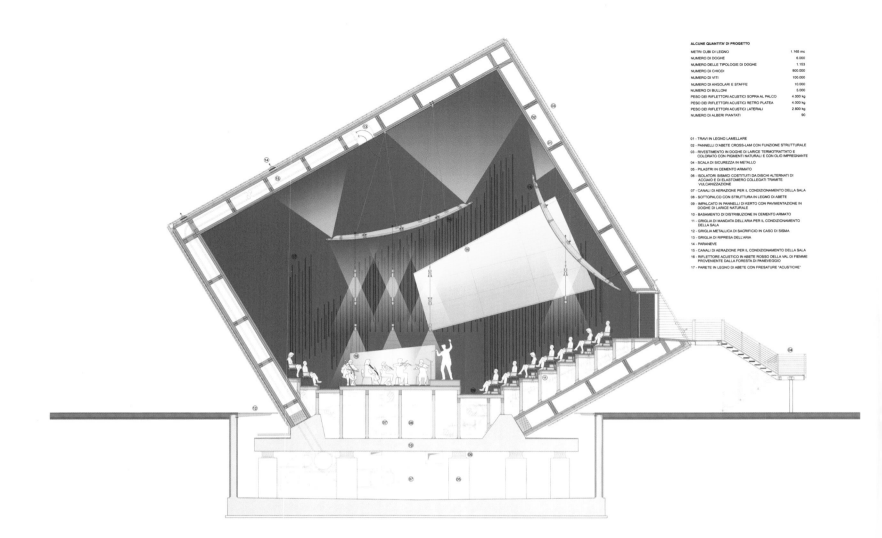

ALCUNE QUANTITA' DI PROGETTO

METRI CUBI DI LEGNO	1.165 mc
NUMERO DI DOGHE	6.000
NUMERO DELLE TIPOLOGIE DI DOGHE	1.153
NUMERO DI CHIODI	800.000
NUMERO DI VITI	100.000
NUMERO DI ANGOLARI E STAFFE	10.000
NUMERO DI BULLONI	5.000
PESO DEI RIFLETTORI ACUSTICI SOPRA AL PALCO	4.000 kg
PESO DEI RIFLETTORI ACUSTICI RETRO PLATEA	4.000 kg
PESO DEI RIFLETTORI ACUSTICI LATERALI	2.800 kg
NUMERO DI ALBERI PIANTATI	90

01 - TRAVI IN LEGNO LAMELLARE
02 - PANNELLI D'ABETE CROSS-LAM CON FUNZIONE STRUTTURALE
03 - RIVESTIMENTO IN DOGHE DI LARICE TERMOTRATTATO E COLORATO CON PIGMENTI NATURALI E CON OLIO IMPREGNANTE
04 - SCALA DI SICUREZZA IN METALLO
05 - PILASTRI IN CEMENTO ARMATO
06 - ISOLATORI SISMICI COSTITUITI DA DISCHI ALTERNATI DI ACCIAIO E DI ELASTOMERO COLLEGATI TRAMITE VULCANIZZAZIONE
07 - CANALI DI AERAZIONE PER IL CONDIZIONAMENTO DELLA SALA
08 - SOTTOPALCO CON STRUTTURA IN LEGNO DI ABETE
09 - IMPALCATO IN PANNELLI DI KERTO CON PAVIMENTAZIONE IN DOGHE DI LARICE NATURALE
10 - BASAMENTO DI DISTRIBUZIONE IN CEMENTO ARMATO
11 - GRIGLIA DI MANDATA DELL'ARIA PER IL CONDIZIONAMENTO DELLA SALA
12 - GRIGLIA METALLICA DI SACRIFICIO IN CASO DI SISMA
13 - GRIGLIA DI RIPRESA DELL'ARIA
14 - PARANEVE
15 - CANALI DI AERAZIONE PER IL CONDIZIONAMENTO DELLA SALA
16 - RIFLETTORE ACUSTICO IN ABETE ROSSO DELLA VAL DI FIEMME PROVENIENTE DALLA FORESTA DI PANEVEGGIO
17 - PARETE IN LEGNO DI ABETE CON FRESATURE "ACUSTICHE"

花园工作室

Garden Studio

Building an extension is not always the solution if you want more living space. At least that was the out of the box answer cc-studio came up with after having been asked to add more space to the ground floor of a family apartment in Watergraafsmeer, a lovely living district of Amsterdam. Extending would have added square metres but would have meant destroying a relatively new kitchen, not adding any new rooms due to the building layout and planning restrictions. But more importantly it would have narrowed the 12m deep garden.

Having a garden in Amsterdam is a rare commodity, having a 10m wide garden is even more special as most real estates are usually only 5 to 6m wide. Strangely this luxury is also problematic as the depth to width ratio (now 12m x 10.5m) already makes the garden feel more wide than deep due to perspective distortion.

Reducing the depth by another 3m would have simply turned the garden into a back yard, leaving little space over for green. Even more so as the two corners of the garden were already occupied by rundown garden sheds.

cc-studio aimed at reinforcing the potential qualities by joining the two volumes into one the resulting the space can be used more effectively. The owner chose to keep part as storage and the rest as a garden studio space.

To strengthen the visual dynamic of the garden, the main diagonal seen from within the living space inside (bottom left) and that runs to the sun lit (top right) back corner is kept un-built. By placing mirrored planes the visual depth of the garden was enhanced to look deeper.

　　如果你想要更多的居住空间，不一定要扩建，至少 cc 工作室在处理阿姆斯特丹美丽的生活区 Watergraafsmeer 的家庭公寓的改建时就是这样。扩建将增加使用面积，但意味着要毁掉一个相对较新的厨房，因建筑布局和规划限制而不能增加任何新的房间，更重要的是，它将把 12 m 深的花园变窄。

　　在阿姆斯特丹拥有一个花园是极为罕见的，拥有一个 10 m 宽的花园更是尤为难得了，因为大多数房产通常只有 5~6 m 宽的花园。奇怪的是这个豪华花园也有长宽比例的问题（现有规格是 12 m×10.5 m），由于透视变形让花园看起来是更宽而不是更长。将长度再减少 3 m 只会把花园变成后院，仅剩下一点绿化空间，甚至花园的两个角落已经被破旧的花园棚占用。

　　cc 工作室旨在增强潜在质量，通过将两个建筑合二而一，使空间可以更有效地被使用。业主选择将部分空间作为存储室，其余为花园工作室。

　　要加强花园的视觉动态效果，从生活空间里（左下）看到的主对角线和朝向太阳的（右上）背角处保持结构一致，通过放置镜面平板，花园的长度在视觉上得到增强，看起来更长。

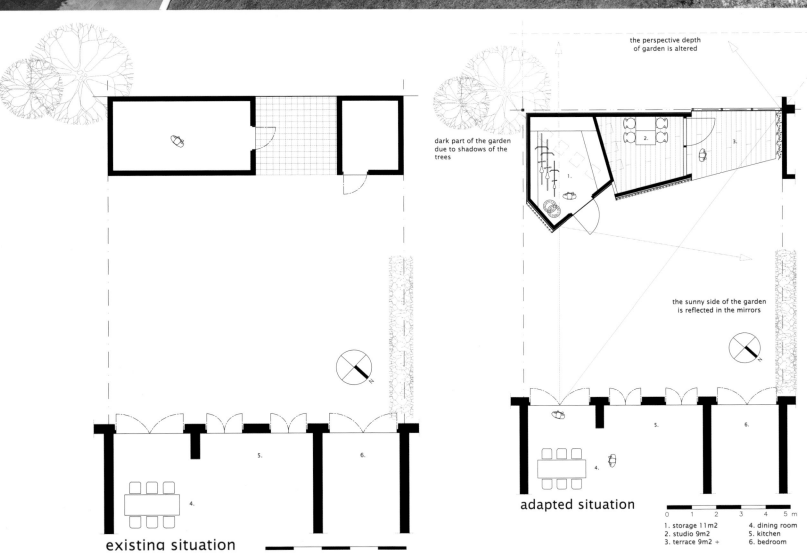

the perspective depth
of garden is altered

dark part of the garden
due to shadows of the
trees

the sunny side of the garden
is reflected in the mirrors

adapted situation

existing situation

0 1 2 3 4 5 m

1. storage 11m2   4. dining room
2. studio 9m2   5. kitchen
3. terrace 9m2 +   6. bedroom

In the corner in which the volume is placed, there is much less sunlight due to the overhanging neighbouring trees. To get more light to that part of the garden, again a mirrored plane was used, this time to reflect the light of the better illuminated section of the garden back. The final measure taken was to make sure that the rebuilt volume would be part of the garden instead of part of the built environment. This was done by covering it with green sedum on faceted angular surfaces to avoid the typical box shape associated to buildings and by hiding the windows from direct sight. This window orientation also gives the studio a sense of intimacy and privacy. So much so that it is not only used as an office space but incidentally doubles as a curtain-less guest room.

The sedum roofing and the hygrical behaviour of the vapour permeable cellulose insulation and gypsum plastered walls create within the critically small space an agreeable indoor climate temperature and healthy air. Also attention was given to create good natural daylight conditions even in the storage part, which is also insulated and can someday, if needed, be added to the studio space.

在花园的角落，由于临近突出树木的遮挡，采光非常少。为了让花园的那一部分得到更多的阳光，再次使用了镜面平面，为花园背部反射更好质量的光线。最后的措施是保证重建部分成为花园的一部分，而不是建筑环境的一部分。这是由用绿景覆盖在方位角表面以避免建筑的典型箱形和避免光线直射窗户。窗口的朝向也给工作室带来亲密感和隐私感，这样工作室不仅可以作为一个办公空间使用，偶尔也可作为客房使用。

绿景天屋顶和蒸汽渗透纤维隔离体及石膏抹灰墙的渗透作用在极小的空间内创造了适宜的室内气候温度和良好空气。同时注意在存储空间也创造良好的自然采光条件，并作隔离处理，如有需要，也可作为工作空间使用。

Grüningen 植物园温室

Greenhouse at Grüningen Botanical Garden

The greenhouse was accomplished in June 2012. It is used as a representative showroom for subtropical plantation.

The new pavilion at the botanical garden in Grüeningen relates strongly to its context. The design was inspired by the surrounding forest, not the built environment. Both the formal vocabulary and the structural concept derive from nature. The pavilion is conceived to harmonize with and expand the forest. The form was developed using Voronoi tessellation, also known as natural neighbor interpolation. Analogous to cell division in nature, the geometry of the roof as surrounding membrane was determined by the position of the old and new trunks. The forest was augmented by four steel trees that form the primary structural system of the pavilion. At about five meters, the trunks branch toward the treetop, which forms the natural roof. A secondary glass construction, suspended from the steel branches, encloses the inner space of the greenhouse.

这座温室于 2012 年 6 月完工，是一个标志性的亚热带种植园展馆。

位于 Grüeningen 植物园的新展馆与其背景有着紧密的联系。该设计灵感来自周边的森林，而不是现成的环境，设计理念和结构概念均源于大自然。这个展馆与森林相称并扩大了森林面积。利用了沃罗诺伊棋盘式布局的形式，这种形式也被称为自然相邻近插值。在性质上类似细胞分裂，作为包膜的屋顶的几何形状由新旧树干的位置决定。构成展馆主要结构系统的四棵钢制树实现了森林的扩展。约五 m 处，树干分支向树梢伸展，形成天然的屋顶。第二个从钢枝悬吊下来的玻璃结构，围绕着温室的内部空间。

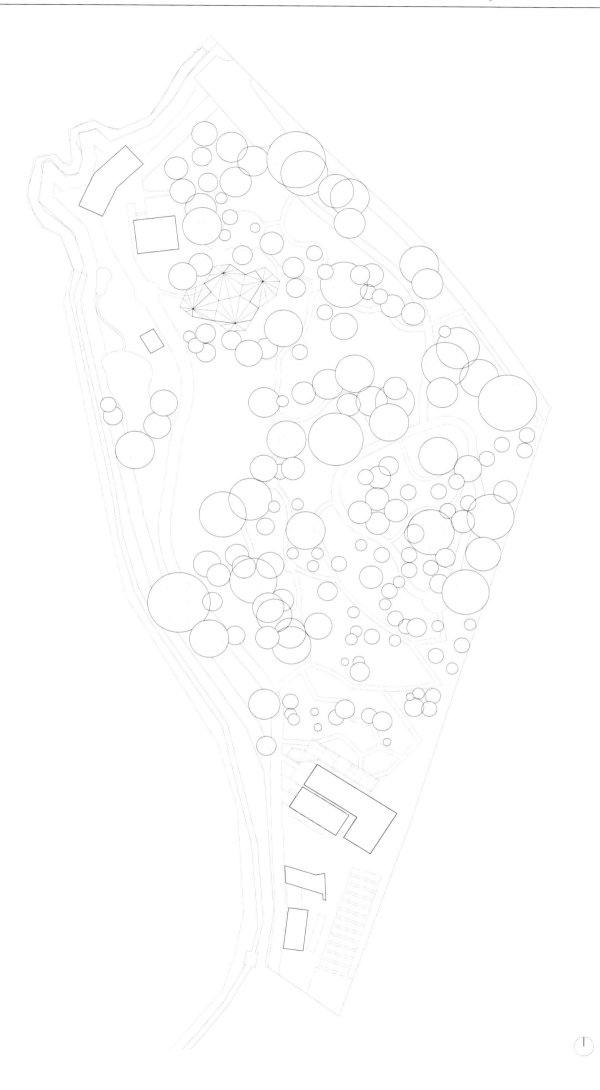

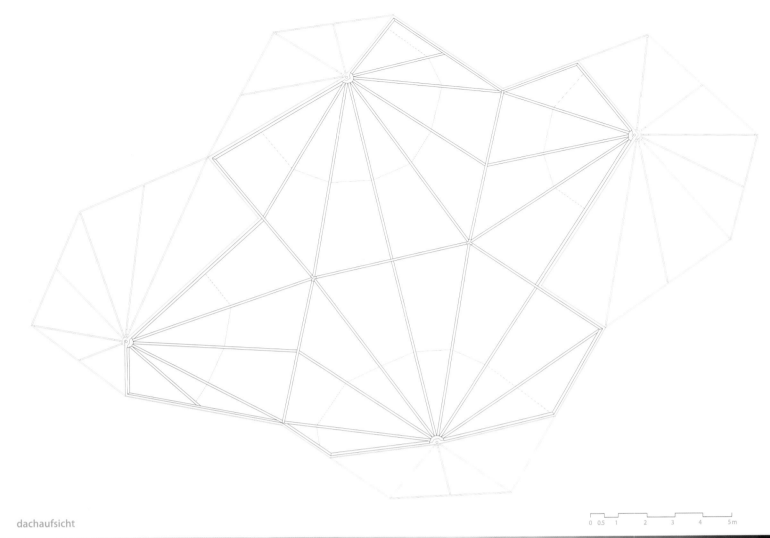

dachaufsicht

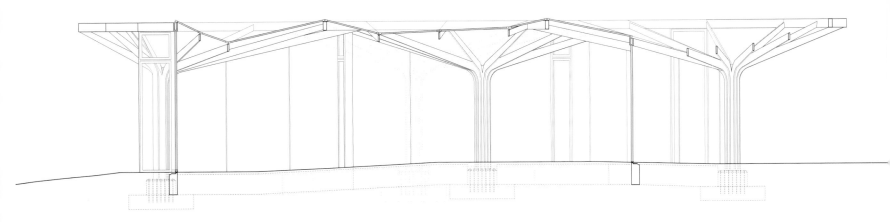

schnitt   0

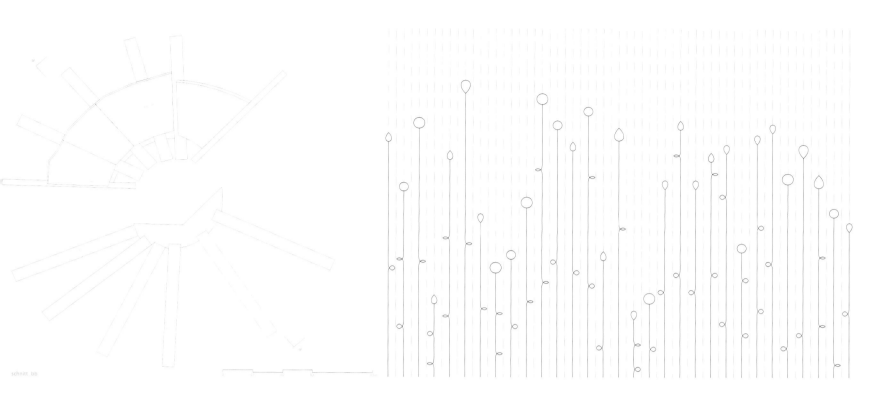

schnitt_bb

schnitt_cc

detail . schnitt

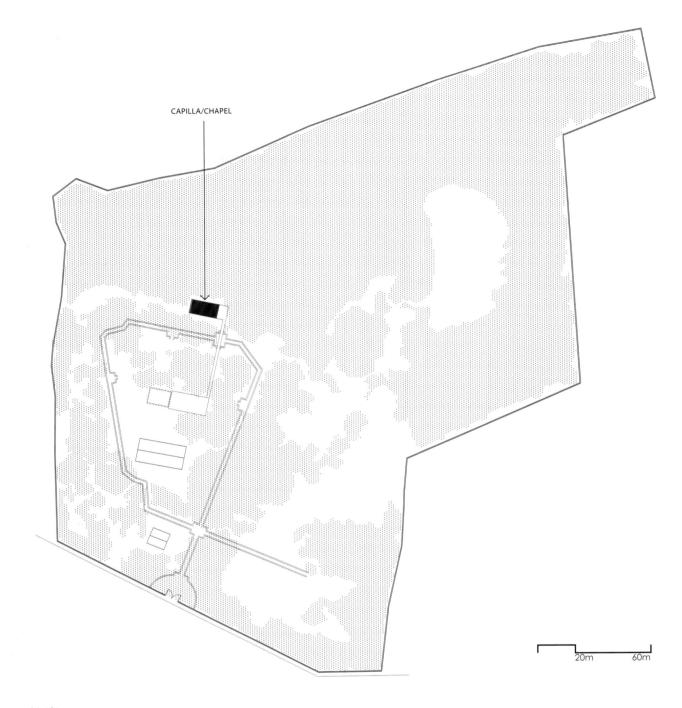

CAPILLA/CHAPEL

20m   60m

La Estancia 教堂
La Estancia Chapel

La Estancia Wedding Gardens were conceived in a traditional Mexican baroque colonial style. The client brief was pretty simple: a colonial-style closed-wall masonry chapel that blended with the surrounding architecture. This deeply troubled us... First of all we did not believe in styles and second, it would be a shame to close the chapel to the surrounding beautiful garden. So we decided to do the complete opposite: an open glass chapel that contrasted with its surroundings.

The site for the chapel was carefully chosen within an enormous area of abundant

vegetation. We selected a location that would not require the removal of any of the existing plants or trees, under large jacarandas which form a natural arch over the chapel and provide it with ample shade.

We believe Tadao Ando's Chapel of Light is a cornerstone in the conception of modern chapels. Around the time of the commission, Steven Holl's Nelson-Atkins Museum of Art had been recently inaugurated and we were deeply impressed by it. We wanted a Tadao Ando meets Steven Holl chapel.

　　La Estancia 婚礼花园曾被构想成传统的墨西哥巴洛克殖民地风格。客户的要求很简单：一个与周围建筑融合在一起的有殖民地风格的封闭石墙教堂。这曾深深地困扰着我们……首先我们不相信风格；其次，将教堂和周围美丽的花园隔开是非常可惜的。所以我们决定设计一个完全相反的风格：一个与周围环境形成对比的开放式玻璃教堂。

　　教堂位置被精心挑选在一个植被丰富的广阔地区，我们选择了一个不需要移

除任何植物或者树木的位置，大型紫薇花树在教堂上方形成了一道的天然拱门，为教堂提供了足够的遮荫。

　　我们认为安藤忠雄的光教堂是现代教堂的奠基石。在设计此项目期间，史蒂芬·霍尔的纳尔逊－阿特金斯艺术博物馆也刚刚完成，给我们留下了极深的印象，于是我们想要设计一个融合安藤忠雄和史蒂芬·霍尔风格的教堂。

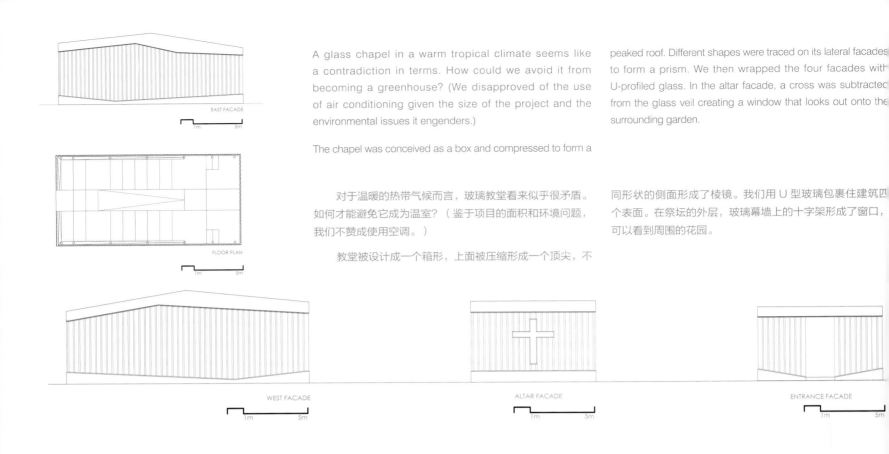

EAST FACADE

1m   5m

FLOOR PLAN

1m   5m

WEST FACADE

1m   5m

ALTAR FACADE

1m   5m

ENTRANCE FACADE

1m   5m

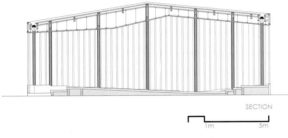

SECTION

1m   5m

A glass chapel in a warm tropical climate seems like a contradiction in terms. How could we avoid it from becoming a greenhouse? (We disapproved of the use of air conditioning given the size of the project and the environmental issues it engenders.)

The chapel was conceived as a box and compressed to form a peaked roof. Different shapes were traced on its lateral facades to form a prism. We then wrapped the four facades with U-profiled glass. In the altar facade, a cross was subtracted from the glass veil creating a window that looks out onto the surrounding garden.

对于温暖的热带气候而言，玻璃教堂看来似乎很矛盾。如何才能避免它成为温室？（鉴于项目的面积和环境问题，我们不赞成使用空调。）

教堂被设计成一个箱形，上面被压缩形成一个顶尖，不同形状的侧面形成了棱镜。我们用 U 型玻璃包裹住建筑四个表面。在祭坛的外层，玻璃幕墙上的十字架形成了窗口，可以看到周围的花园。

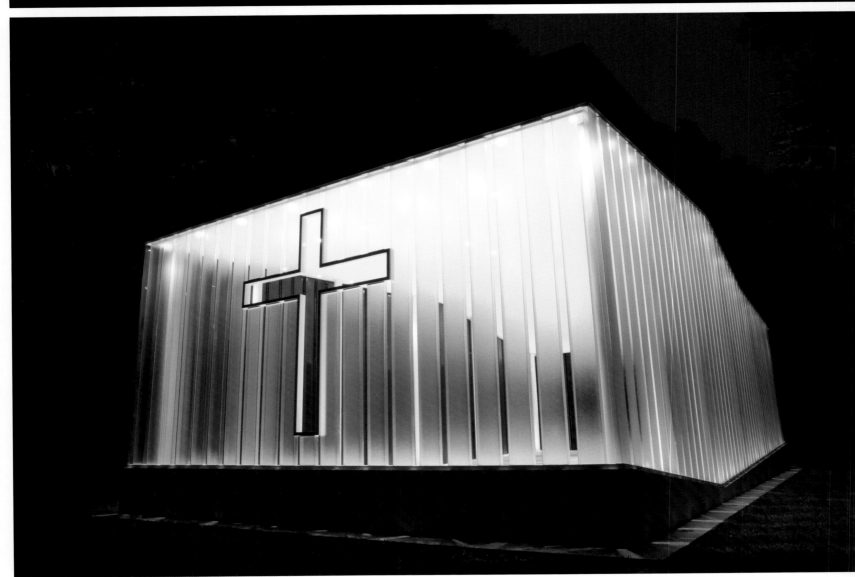

日落教堂
Sunset Chapel

Our first religious commission, La Estancia Chapel, was a wedding chapel conceived to celebrate the first day of a couple's new life. Our second religious commission had a diametrically opposite purpose: to mourn the passing of loved ones. This premise was the main driving force behind the design. The two had to be complete opposites, they were natural antagonists. While the former praised life, the latter grieved death. Through this game of contrasts all the decisions were made: glass vs. concrete, transparency vs. solidity, ethereal vs. heavy, classical proportions vs. apparent chaos, vulnerable vs. indestructible, ephemeral vs. lasting…

The client brief was pretty simple: first, the chapel had to take full advantage of the spectacular views; second, the sun had to set exactly behind the altar cross (of course, this is only possible twice a year at the equinoxes); and last but not least, a section with the first phase of crypts had to be included outside and around the chapel. Metaphorically speaking, the mausoleum would be in perfect utopian synchrony with a celestial cycle of continuous renovation.

　　我们的第一个宗教设计任务，La Estancia 教堂，是一个庆祝夫妻新生活第一天的婚礼教堂。第二个宗教任务有着截然相反的目标：缅怀逝去的亲人。这个前提是设计背后的主要驱动力。两个任务必须完全对立，它们是天生的对立者，前者赞美生命，后者哀悼死亡。通过这种对比所得出的结论是：玻璃对混凝土，透明对实心，轻盈对沉重，古典均匀对明显混乱，脆弱对坚不可摧，短暂对持久……

　　客户的要求很简单：首先，教堂要充分利用壮观秀丽的景色；其次，太阳要精确设置在祭坛后面的交叉处（当然，一年只有春秋分时才可能出现 2 次）；最后但同样重要的是，教堂地下室第一部分要在户外并且围绕着教堂。这意喻着陵墓将与天空的周而复始完美同步。

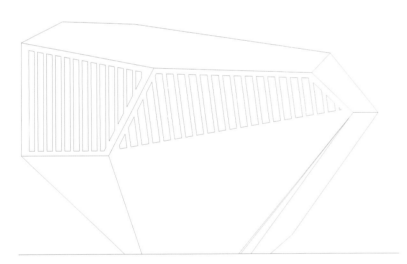

01　ELEVACION SUR

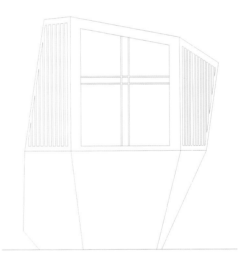

02　ELEVACION PONIENTE

03　ELEVACION NORTE

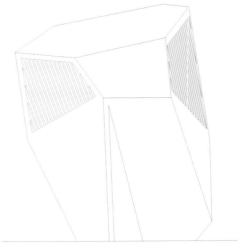

04　ELEVACION ORIENTE

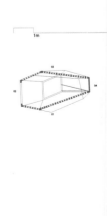

A5.01

ELEVACIONES

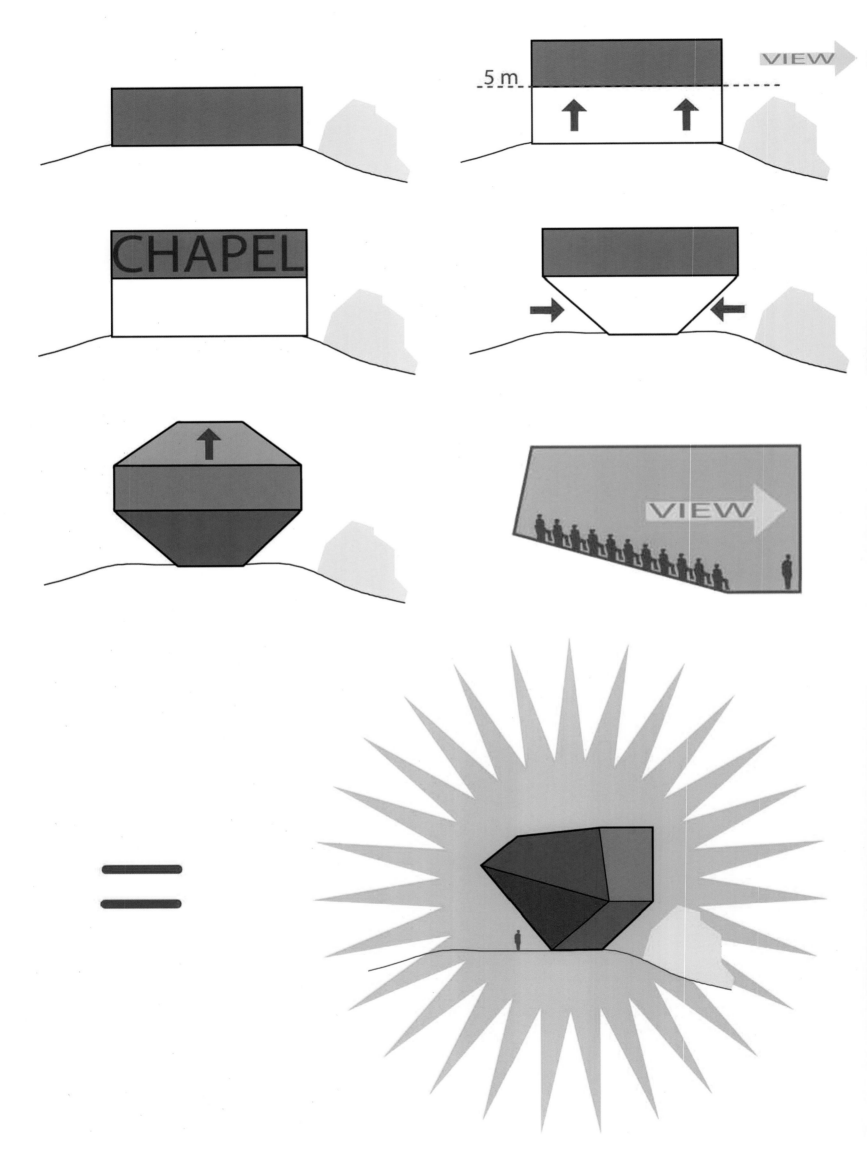

CHAPEL

5 m

VIEW

VIEW

=

Two elements obstructed the principal views: large trees and abundant vegetation, and a behemoth of a boulder blocking the main sight of the sunset. In order to clear these obstructions (blowing up the gigantic rock was absolutely out of the question for ethical, spiritual, environmental and, yes, economical reasons), the level of the chapel had to be raised at least 5 m. Since only exotic and picturesque vegetation surrounds this virgin oasis, we strived to make the least possible impact on the site, reducing the footprint of the building to nearly half the floor area of the upper level.

Acapulco's hills are made up of huge granite rocks piled on top of each other. In a purely mimetic endeavor, we worked hard to make the chapel look like "just another" colossal boulder atop the mountain.

两个因素阻挡了主要景色：大型树木和茂盛的植被，以及庞大的巨石挡住了夕阳的光线。为了清除这些障碍（炸掉巨石是绝对不行的，不管是从道德、精神、环境还是经济理由上），教堂的水平位置要提高至少五米。由于异国情调和风景如画的植被包围着这片原始绿洲，我们力求把对此地的影响降到最小，将建筑物将近一半面积分到上层。

阿卡普尔科的群山由巨大的花岗岩石块堆积构成。在一个纯粹的模仿尝试中，我们努力让教堂看起来像一块在山峰上的巨石。

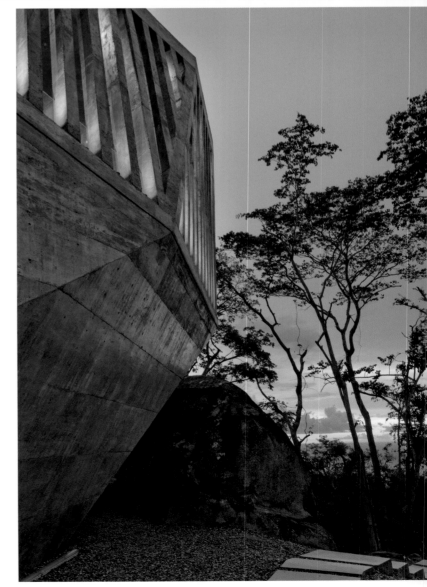

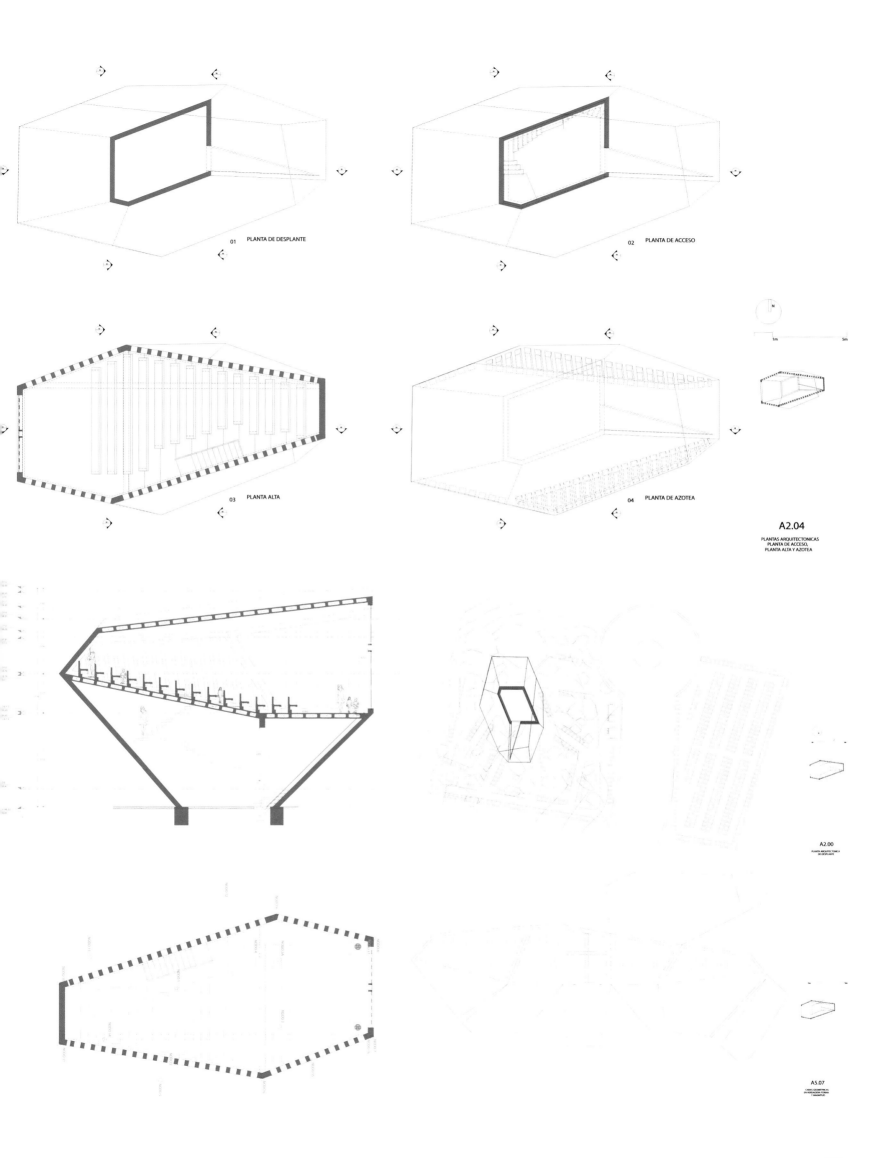

01 PLANTA DE DESPLANTE

02 PLANTA DE ACCESO

03 PLANTA ALTA

04 PLANTA DE AZOTEA

N

1m 5m

A2.04

PLANTAS ARQUITECTONICAS
PLANTA DE ACCESO,
PLANTA ALTA Y AZOTEA

A2.00

PLANTA ARQUITECTONICA
DE DESPLANTE

A5.07

FORMAS GEOMETRICAS
EN VERDADERA FORMA
Y MAGNITUD

3E——弗罗茨瓦夫理工大学研究与教育综合大楼

3E – Research & Educational Complex Wroclaw University of Technology

The "3E" (Energy, Ecology, Education) Research and Educational Complex of the Faculty of Environmental Engineering will be the most progressive and sustainable combination of buildings on a campus designed for the Wroclaw University of Technology in Wroclaw (Poland).

The main building of the complex is a zero-energy educational building with lecture auditorium, seminar rooms and laboratories. Additionally, building will operate as a permanent exhibition where various modern green technologies, facilities and materials will be on show. It will be also a kind of field research laboratory for building energy efficiency and use of renewable energy.

Its architecture is originated in energy and ecological requirements. The form of this high-performance building is designed to optimise the

use of solar energy and to reduce heat losses during cold periods and overheating in summer. All construction materials will be environmentally safe, most of them will be recyclable and absolutely safe in terms of human health hazards. The energy for these buildings is collected from sun, ground and water. Building systems will be constantly monitored.

The design of the "3E" complex is a result of a close multidisciplinary cooperation between architect and the team of scientists from university's Faculty of Environmental Engineering. As educational buildings they will be a flagship project that will propagate throughout the country the idea of energy saving and ecological architecture.

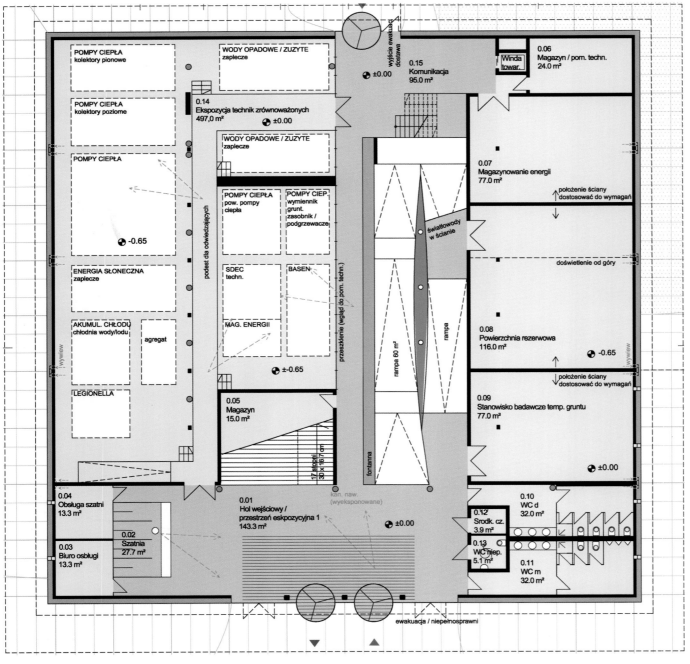

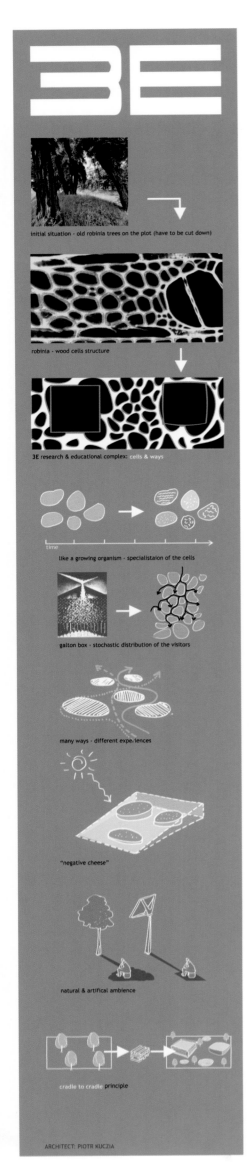

initial situation - old robinia trees on the plot (have to be cut down)

robinia - wood cells structure

3E research & educational complex: cells & ways

like a growing organism - specialistaion of the cells

galton box - stochastic distribution of the visitors

many ways - different experiences

"negative cheese"

natural & artificial ambience

cradle to cradle principle

ARCHITECT: PIOTR KUCZIA

ENERGY ECONOMY ECOLOGY
RESEARCH AND EDUCATIONAL COMPLEX OF THE FACULTY OF ENVIRONMENTAL ENGINEERING
NEW CAMPUS - UNIVERSITY OF TECHNOLOGY WROCLAW | POLAND

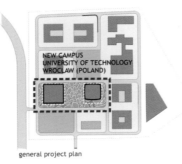

NEW CAMPUS
UNIVERSITY OF TECHNOLOGY
WROCLAW (POLAND)

general project plan

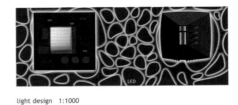

light design 1:1000

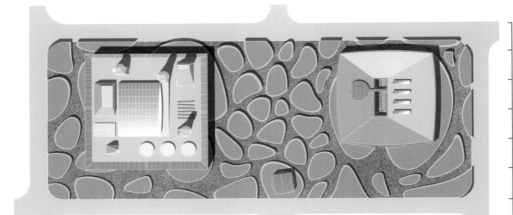

phase ONE 1:500

phase X 1:500

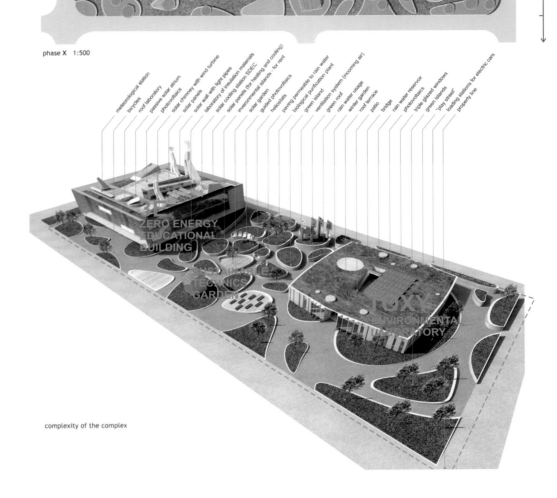

complexity of the complex

弗罗茨瓦夫理工大学环境工程学院"3E"（能源、生态、教育）研究与教育综合大楼将成为校园里最先进的可持续性建筑综合体。

综合大楼的主建筑是零能源教学楼，包括演讲厅、会议室和实验室。此外，大楼将作为一个永久性的展览空间，展示各种现代环保技术、设备和材料。它也将是一种建筑能效研究和再生能源应用的研究实验室。

它的建筑设计起源于能源和生态需求。这种高性能建筑被设计成优化太阳能的

利用以及减少在冬季的热量损失和夏季过热的形式。所有的建筑材料都是环保的，它们中的大多数都是可回收的，并且不会危害人类健康，绝对安全。这些建筑的能量来自阳光、地面和水。建筑系统将处于持续监测中。

"3E"综合大楼是建筑师和环境工程学院的科学家小组之间的密切合作的复杂结果。作为教育建筑，它们将成为节能和生态建筑的旗舰项目，将这种理念传播到整个国家。

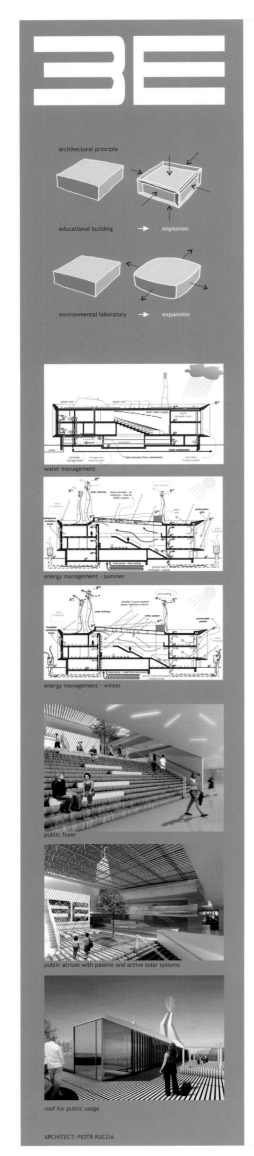

architectural principle

educational building → implosion

environmental laboratory → expansion

water management

energy management - summer

energy management - winter

public foyer

public atrium with passive and active solar systems

roof for public usage

ARCHITECT: PIOTR KUCZIA

cells & artifacts

islands scenery

plaza

urban cell structure

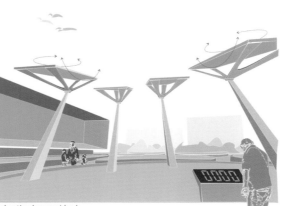

educational energy island

footpath

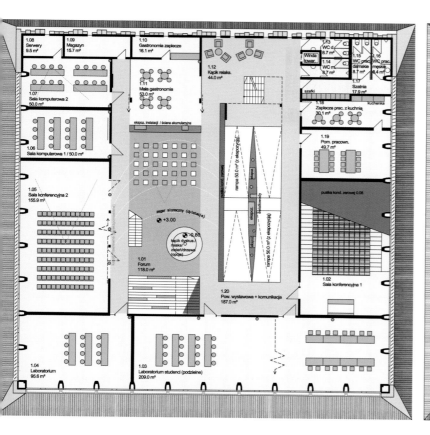

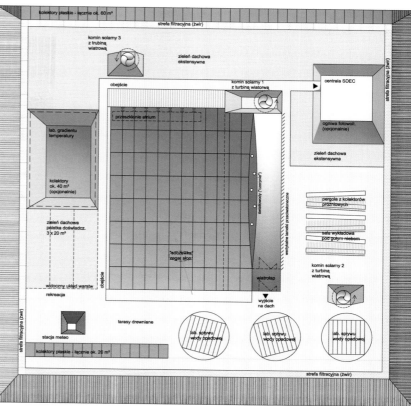

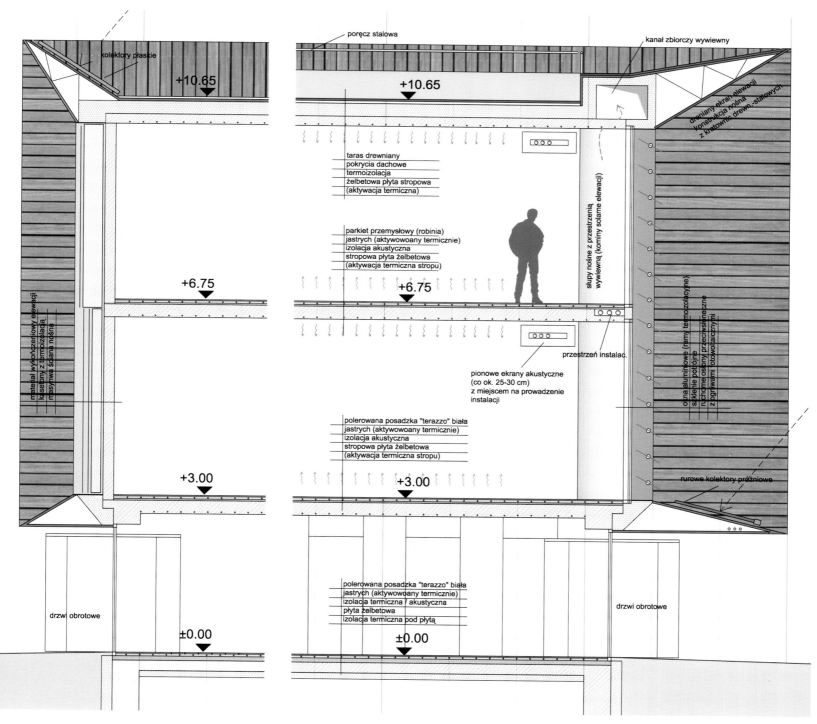

中国昆山采摘亭

Harvest Pavilion

Sited in an eco-farm alongside the Yang Cheng Lake, Kunshan, the project consists of 4 small scale public buildings: a club house, a harvest pavilion, a botanical showroom, and an information center. In the spring of 2012, the harvest pavilion became the first one completed.

The farm is vast, flat, and wide open to sky. Different from the congested vertical massing image of urban life, such an empty flatness of the site is an essential nature that we believe the architecture should respond to. Our design task is to explore how architecture should be integrated into such a context, to create a new and unique place, however harmonize with nature.

The harvest pavilion appears a simple, light, and translucent cuboid, with a horizontal thin plane hovering at the top, flying parallel with the horizon in the distance. The plane, made of pre-fabricated aluminum rods, cantilevers out at 4 sides at various depths. The space below becomes a transition zone from the interior to exterior, and promotes the potential activities because of the pleasant shadow casted by the canopy. The building facade system consists of vertical laminated bamboo louvers, floor-to-ceiling frameless glass panels, and pivoting glass doors. The transparency and lightness of such a material combination visually fuse the building volume with its surrounding landscape, and make the architecture sensitive to light. Under the condition of nice weather, when the pivoting glass doors are all rotated to open, the indoor space is literally stretched out into the farmland.

　　该项目位于昆山阳澄湖的一个生态农场，农场 由4个小型公共建筑构成，会所、采摘亭、植物展厅和信息中心。2012春，采摘亭项目第一个建成。

　　这座广阔平坦的农场直通天边。不同于都市生活中的楼宇密布的拥挤形象，场地的空旷平坦是一个重要特质，我们认为建筑设计应该呼应这种特质。我们的设计任务是探讨建筑如何融入到这样的环境中，创建一个新颖独特的场所，而且还能与大自然相协调。

　　采摘亭看似简单、轻盈、半透明的长方体，水平而狭长的平面盘旋于顶部，与远处的地平线平行延伸。这个平面由预制铝棒做成，从四个方向悬挑出去，伸出的长度各不相同。顶篷下方的空间成为一个由内至外的过渡区，起到了遮阳的作用，提供舒适的环境，促进了各种活动的举办。建筑立面系统由垂直层压竹制的百叶、通高的无框玻璃板和旋转玻璃门构成。这样的材料组合所具有的透明度和轻盈度在视觉上建筑体量与周围景观融为一体，并使建筑对光线敏感。在天气宜人的情况下，玻璃旋转门全部打开，室内空间实际上一直延伸到农田里。

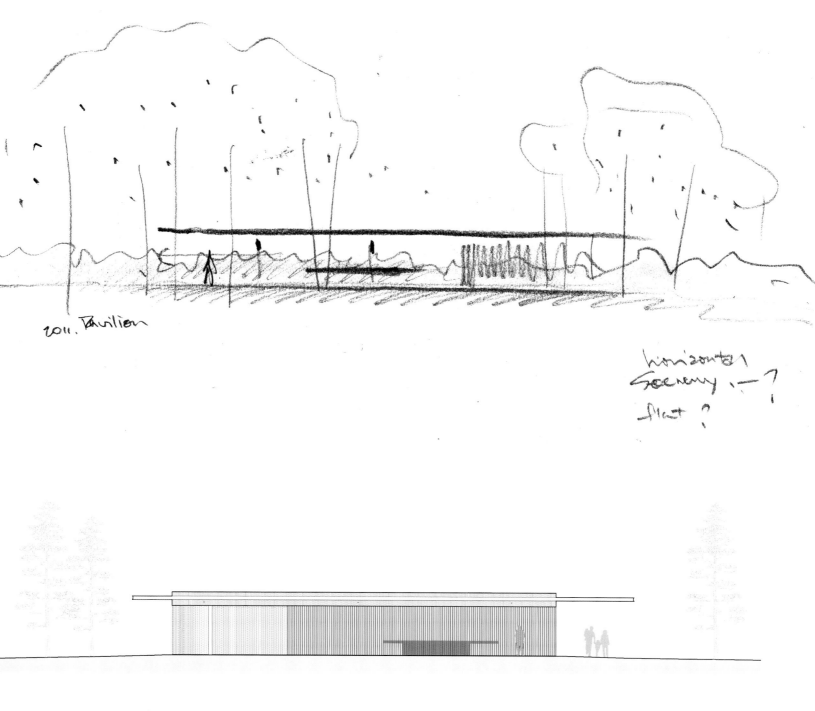

2011. Pavilion

horizontal
scenery : ─?
flat ?

0m 1m 2m   4m

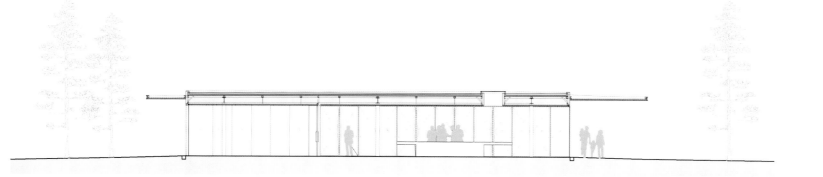

0m 1m 2m   4m

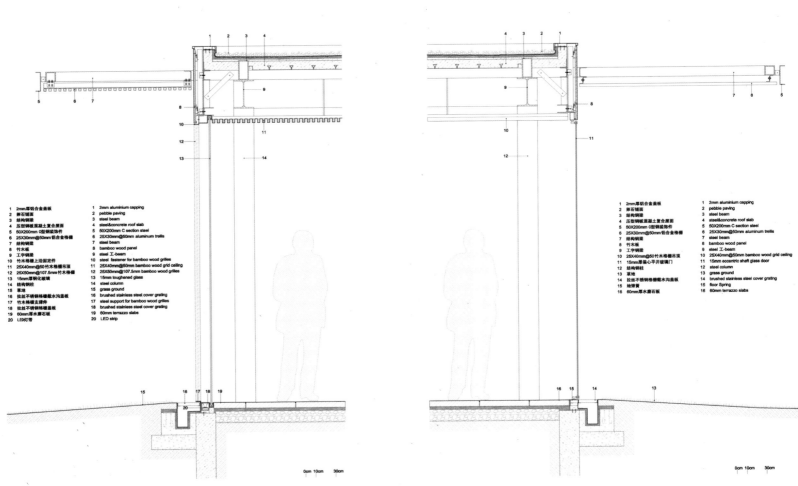

1 2mm厚铝合金盖板	1 2mm aluminium capping
2 卵石铺面	2 pebble paving
3 结构钢梁	3 steel beam
4 压型钢底混凝土复合屋面	4 steel&concrete roof slab
5 50X200mm C型钢装饰件	5 50X200mm C section steel
6 25X30mm@50mm铝合金格栅	6 25X30mm@50mm aluminum trellis
7 结构钢梁	7 steel beam
8 竹木板	8 bamboo wood panel
9 工字钢梁	9 steel 工-beam
10 竹木格栅上沿固定件	10 steel fastener for bamboo wood grilles
11 25X40mm@50竹木格栅吊顶	11 25X40mm@50mm bamboo wood grid ceiling
12 25X50mm@107.5mm竹木格栅	12 25X50mm@107.5mm bamboo wood grilles
13 15mm厚钢化玻璃	13 15mm toughened glass
14 结构钢柱	14 steel column
15 草地	15 grass ground
16 拉丝不锈钢格栅截水沟盖板	16 brushed stainless steel cover grating
17 竹木格栅支撑件	17 steel support for bamboo wood grilles
18 拉丝不锈钢格栅盖板	18 brushed stainless steel cover grating
19 60mm厚水磨石板	19 60mm terrazzo slabs
20 LED灯带	20 LED strip

1 2mm厚铝合金盖板	1 2mm aluminium capping
2 卵石铺面	2 pebble paving
3 结构钢梁	3 steel beam
4 压型钢底混凝土复合屋面	4 steel&concrete roof slab
5 50X200mm C型钢装饰件	5 50X200mm C section steel
6 25X30mm@50mm铝合金格栅	6 25X30mm@50mm aluminum trellis
7 结构钢梁	7 steel beam
8 竹木板	8 bamboo wood panel
9 工字钢梁	9 steel 工-beam
10 25X40mm@50竹木格栅吊顶	10 25X40mm@50mm bamboo wood grid ceiling
11 15mm厚偏心平开玻璃门	11 15mm eccentric shaft glass door
12 结构钢柱	12 steel column
13 草地	13 grass ground
14 拉丝不锈钢格栅截水沟盖板	14 brushed stainless steel cover grating
15 地弹簧	15 floor Spring
16 60mm厚水磨石板	16 60mm terrazzo slabs

神乐坂琥珀餐厅

Kagurasaka Kohaku

The restaurant Kohaku is situated in "Kagurazaka", a very traditional district in Tokyo. It is on the first floor of an existing building. The approach to the restaurant is through a small, inner garden surrounded by a black bamboo wall from the street, after which a small pond greets the guest. This is in the image of a traditional Japanese art form called "Bon-kei".

By opening the wooden lattice door, one enters into a space of black mud walls with a gentle light leaking through the wood grill ceiling. Here, the "tokonoma" (alcove) image, an element of Japanese architecture, has been developed into a small gallery. Facing this entry, is a "kake-juku" (scroll) expressing the present season and, which is changed with each season of the year.

In the restaurant area, the black mud walls change to white, and various patterns of latticed woodwork in the "shoji" (a traditional partition using washi paper) together with "sumi" drawings on washi paper appear as elements composing the entire space. Furthermore, elements such as black lacquer panels (acrylic panels with lacquer on powdered charcoal) and "karakami" paper (paper with traditional patterns by wood block printing) representing some of Japan's traditional techniques, are expressed in a contemporary space.

At this restaurant, one is able to enjoy a space where the traditional and the modern mingle in harmony.

　　琥珀餐厅位于东京一个非常传统的地区 ——"神乐坂"，在一幢已有建筑的的一层。通过街上黑色竹墙包围的一个小室内花园进入餐厅，一个小池塘在后面欢迎着客人，这是被称为"Bon-kei"的传统日本艺术形象。

　　打开木格子门，便进入到一个黑泥壁房间，柔和光线从木格栅天花板中倾泻下来。在这里，日本建筑元素的"壁龛"，已经变为一个小画廊。正对入口处，是一个表示季节且随着季节变化的"卷轴"。

　　餐厅区，黑泥墙壁变为白色，各种格子图案的"商纸"（传统隔墙用纸）与在商纸上的"sumi"图案一起构成完整的空间元素。此外，各种元素如黑漆板（木炭粉漆亚克力板）和"karakami"纸（传统用木版印刷图案的纸张）代表了一些日本传统技术，在当代空间中得到表现。

　　在这家餐厅里，人们能够享受到传统与现代和谐交融的空间。

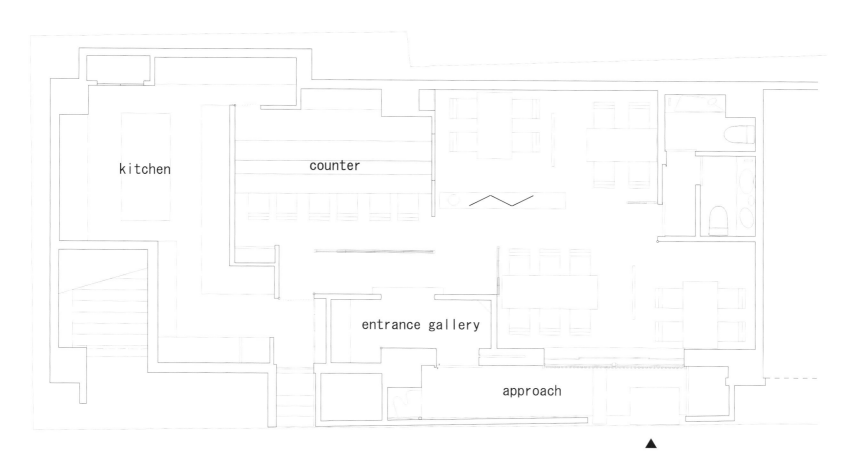

kitchen

counter

entrance gallery

approach

虎白平面図　S=1/100

Contributors / 设计师名录

Archivision Hirotani Studio

YOSHIHIRO HIROTANI / Architect
1956 Born in Wakayama Prefecture, Japan
1980 Graduated from Tokyo University of Science
Joined Archivision Architect & Associates
2006 Established Archivision Hirotani Studio with YUSAKU ISHIDA

YUSAKU ISHIDA / Architect
1969 Born in Saitama Prefecture, Japan
1994 Graduated from Tokyo City University with Master's Degree in Architecture
1996 Joined Archivision Architect & Associates
2006 Established Archivision Hirotani Studio with YOSHIHIRO HIROTANI

Prize
2008 JCD BEST 100, Osho-chiku Community Center
2009 The prize of Architecture of Toyama Prefecture, Osho-chiku Community Center
2011 Good Design Prize for Leimond Nursery School in Nagahama
2011 Kids Design Award for Leimond Nursery School in Nagahama
JCD Second Prize, Japanese Society of Commercial Space Designers for Leimond Nursery School in Nagahama
Architectural Institute of Japan, Selected Architectural Designs for Leimond Nursery School in Sakado

baumraum

The architectural office baumraum, located in Bremen, Northern Germany, is specialized in the planning of experimental constructions on the ground, in trees and by or on water.

Since 2003 baumraum has planned and built inspiring dwellings for children and adults – from small playhuts for games and activities to exclusive permanent abodes. baumraum specializes in planning and realization of treehouses.

The constructions designed by baumraum are noted for their originality and inspiring creativity. Individual concepts are designed for private clients, hotels and catering businesses, forest environment and for special events. The safety and durability of the constructions are of primary importance.

The main concern thereby is to handle the trees and their surroundings with the utmost care, ensuring their protection and preservation. The treehouses are not anchored by bolts or nails or any other measures, which might injure the trees, but rather by textile belts and adjustable steel cables, harmless to them.

The principal material used is domestic timber with its excellent properties like sustainability, water resistance and colour. But even metal, textiles and plastics are potential materials for the architects. baumraum combines the creative and constructive expertise of architects with the long-standing experience of landscape designers, tree experts, and established, reputable craftsmen.

The architects design suitable projects both in the natural and the urban environment. So far they have realized more than 40 projects in Europe – Germany, Austria, Italy, Czech Republic and Hungary – but also in Brazil and the USA.

Bunker Arquitectura

Bunker Arquitectura is a Mexico City-based architecture, urbanism and research office founded by Esteban Suarez in 2005 and partnered by his brother Sebastian Suarez. In their short career they have been able to experience and experiment architecture in the broadest scale possible: from small iconic chapels for private clients to a master plan for an entire city. Bunker has continuously attracted attention for its unconventional approach to architecture with projects such as a three-kilometer habitable bridge that unites the bay of Acapulco and an inverted skyscraper 300 meters deep in the main square of the historic center of Mexico City.

Every new project starts with a profound research of the social, political, economical, cultural and environmental factors that surround each particular site. The analysis and understanding of all this information, crossbred with Bunker's indefatigable pursuit of innovation, yields architecture that is specific to its conditions. In this sense, no two projects ever look or feel alike. What ties them together is an evident need to constantly push the boundaries of architecture.

Besides developing projects for private clients, the government or competitions, Bunker is continuously involved in self-financed research projects that nurture the theoretical side of their practice. In this manner, the built and unbuilt projects bear the same weight in their balance. Theory and practice coexist in perfect symbiosis.

"STOP: KEEP MOVING: an oxymoronic approach to architecture" is Bunker's first monograph. Their belief that a contradictory view of life, the human condition and architecture is central to finding architectural meaning has led them to rely on the oxymoron, opposite words or ideas that when put together reveal a new meaning, to disclose their creative processes and ingenious solutions to eccentric demands.

Cazu Zegers Arquitecturea

Suggests an undertaking of Chilean architecture, to find the construction of a language of expressive forms and arquitecture intimately connected to Chile, its territory, landscape, and its vernacular construction.

In this sense the works developed by this studio, do not pretend to be a finished work, but a work in progress, where there is poetic reflection, about the way habitation of this territory. They are "Prototypes of this Territory." (Name taken by the monographic book, published by Ediciones ARQ, in 2008). Tierra Patagonia is an excellent example of this artistic and creative posture.

The studio has a very particular design outline, that comes from a relation between poetry and arquitecture, inserted in the thesis of AMEREIDA (published in Ediciones de la Escuela de arquitectura de la UCV), which suggests that the coming of Columbus to the Indies and America appears as a gift, a talent. Amereida invites the inhabitants of America to construct a language of forms appropriate to our Latin heritage in an American world, that is: assume the responsibility as a new culture, the Latin-Americans, and from this origin find a way to dialogue with the worldwide paradigm. In answer to this invitation, Cazu Zegers, calls it "Habitation in an Delicate and Precaurious way," that is to say with low technology or low tech and a high experiential impact. This particular outlook is born from the understanding, that the great heritage of Chile is in the territory, in which:... "Chile before a country is a landscape", as sung by the poet Nicanor Parra.

All his work is construed under the premise: "The territory is to America, as the monuments are to Europe." From here this arquitecture does not set out to be a protagonist but to enter into an amorous dialogue with nature.

cc-studio

Architecture is made for people by people and built within our natural environment. We care about all 3. Our niche is in projects in which we can inventively combine technical and sustainable innovations with strong architectural qualities like remarkable spatial experience and efficient functionality.

We like to work for open minded clients and operate well with in the constraints of tight budgets. Good constraints trigger more "out of the box thinking", more creativity, fun and inventiveness during the whole process. An open mind is important to be able to see and make use of opportunities along the road. As a small studio we believe strongly in an intelligent and interactive creative collaboration with all involved in the process of the design and it's realization: clients, advisors, (sub-) contractors and industries. Our office is situated in a multidisciplinary creative industrial environment.

Our ongoing projects are the construction of a small landscaped and green garden studio in Amsterdam, study for 30 greenhouse covered low-cost, low energy and adaptive housing (financed by the Dutch Innovation funds Agentschap NL and SEV foundation for experimental housing research in collaboration with the Technical University of Eindhoven) .

Diébédo Francis Kéré

As is the tradition in Burkina Faso, each member of the family is responsible for the well-being of all members. Each individual is indispensable for the survival of the community. If one member leaves the community in search of a better life, he tries to compensate his loss by sending back financial support.

Francis Kéré wishes to fulfill his part of the social duty by offering a new perspective to the whole village. His projects reach way beyond simple financial support. His various stays in Europe have shown him that education and training are the basis of any social, professional and economic development. For this reason, our first goal was to build a school in Gando, providing education to many children. Since then, several development projects have been set up to improve the living conditions in the village.

Helping people to help themselves constitutes the basis of all our projects. According to our philosophy, long-term considerations are the priority for proper and sustainable development aid. In our opinion, the key of sustainable development relies on the population's participation. The whole village community is involved in the building process: the men manufacture the clay bricks, lay the foundations, build up the walls and install the roofs; the women beat the clay floors and plaster the walls. Every morning for a whole year, the children bring a stone for the foundations on their way to school. The population's participation integrates the projects into the local communities and enables identification and motivation. For this reason, the results are valued, preserved and further developed.

Every innovation is first introduced into the school and afterwards into the society. Children accept change more easily than adults, and the next generations will naturally accept and use new facilities, for example latrines to prevent intestinal diseases.

We make sure that the constructions are as inexpensive and simple as possible. In order to ensure that the population is able to use similar techniques, the buildings are made of clay which is found everywhere in the region. The extreme climatic conditions of Burkina Faso require a particular and adapted architecture. Our construction method, using natural ventilation to cool down the room climate, is appropriate for the temperature. The overhanging roof, the clay walls and the natural cooling system prevent the building from overheating. The whole country is currently speaking of the benefits of this way of constructing.

Emma Architects, Amsterdam

Emma designs buildings that are constructed with care, integrated in their environment with care, that are beautiful and that are well. Emma produces buildings in and around which users can live a happy and healthy life.

Emma was established in 2005 by partner architects Jurg Hertog and Marten de Jong.

Jurg Hertog leads Emma Create, the engineering and production branche of Emma. Emma Create operates from her studio in Amsterdam with an international team of designers and engineers. Marten de Jong leads Emma Explore, emma's independent think tank on spatial issues.

idA buehrer wuest architekten sia ag

idA is a young design studio founded in 2010 in Zuerich. idA engages in projects of every description. Nevertheless all projects are connected through the studios methode of operation. idAs projects are characterized by an approach intent on context and concept of each building project. It's interest lies in developing very specific solutions for the most diverse tasks.

Stephan Buehrer

Born in 1973 in Zuerich, apprenticeship as architectural draughtsman, Diplom Architect, Project Architect for offices in Zuerich, London and Paris, foundation Stephan Buehrer Architekten in Zuerich from 2004 to 2010, collaboration BBKA, Basel since 2005, Assistance ETH Zuerich for the Chair of Marc Angélil from 2006 to 2007, founder of idA buehrer wuest architekten ag in Zuerich in 2010

Martina Wuest

Born in 1985 in Zuerich, MSc ETH Zuerich, founder of idA buehrer wuest architekten ag in Zuerich in 2010

James & Mau

James & Mau is an international architecture firm based in Madrid that brings together the expertise in several areas such as architecture, construction and real estate development in order to offer a full service support for the development of the projects both for individuals and enterprises.

The company is widely experienced in design, work on-site and building permits, with a strong focus on modular industrialized architecture and construction, applying bioclimatic and sustainable concepts.

One of the added values of James & Mau is its international profile with a strong knowledge of Latin American market, especially of Chile and Colombia where it has its offices. The firm aims to position itself as a platform for European investors overseas.

James & Mau was founded in 2007 by Jaime Gaztelu and Mauricio Galeano, and in turn, participates in Infiniski, company dedicated to modular and sustainable construction very much concerned with high quality design, fast execution and budget accuracy.

Jauncarlos Fernandez

Architect Jauncarlos Fernandez worked hand-in-hand with the CADE winemaking team to create a state-of-the-art winery whose simple, striking design reflects a straightforward, sustainable approach to winemaking. Their shared vision of environmental responsibility helped CADE become the first organically farmed LEED (Leadership in Energy and Environmental Design) Gold Certified estate winery in Napa Valley. Building elements include concrete with integral earth colors, steel made from 98 percent recycled material and hundreds of square feet of glass for a well-lit working environment that conserves energy. "I try to stay away from nonrenewable materials," Fernandez says. "We need our

natural resources to continue building into the future."

After earning his Professional Degree in Architecture from ITESO, a private university in Guadalajara, Fernandez traveled extensively for projects all over the world, some of which include a golf clubhouse in Mexico, a concept house in Japan, a guest house in Millbrook, New York, and a residence for a member of the Mondavi family in Calistoga, California. He has also contributed to the design of several professional buildings and residences in Napa Valley, as well as two other wineries. Fernandez has been a registered architect in Mexico since 1992.

KUCZIA / Peter Kuczia

Born in Poland. Studied architecture at Silesian University of Technology, Poland. 2008 dissertation on solar energy use in architecture. Since 1999 he has worked for agn Group in Ibbenbüren, Germany and as freelancing architect in Poland. He has been

awarded several international prizes. His work – which covers a wide range of building types – has been exhibited and published around the world.

modulorbeat

Marc Günnewig and Jan Kampshoff, born in 1973 in Münster and in 1975 in Rhede, the pair studied Architecture at msa | münster school of architecture. They have worked and taught at several universities, among others the University of Auckland, New Zealand

and at present the University of Kassel. In 1999 they co-founded modulorbeat, a network of architects, urbanists and designers headquartered in Münster, which shot to fame with its temporary experimental structures.

Moriyama & Teshima Architects

ntegrating land, building, people and process for over 50 years. Our built world, and the land and environment it relates to, is the context in which we live, work and play. We believe that our built world reflects and ffects the very essence of who we are, how we value our environment, nd what we aspire to for ourselves and for our collective future.

Moriyama & Teshima (M&T) has been delivering high-quality, ustainable and award-winning architecture, landscape architecture nd urban and environmental planning and design services for over 0 years, and is particularly engaged by community centered projects nd "Places for Learning" including museums and galleries, science entres, libraries, cultural and faith facilities, and school, college and niversity buildings and master plans.

ather than merely "sustainable design" M&T designs for a ustainable world and, with renowned landscape architecture / urban environmental planning group integrated into the office, also roduces ground-breaking environmental master plans and designs uch as the "Meewasin Valley 100 Year Master Plan", the "Niagara arks 100 Year Vision, 20 Year Plan and Five Year Action Plan."

Moriyama & Teshima is currently leading the Master Planning of the Holy Cities of Makkah and Madinah, Kingdom of Saudi Arabia, and – as well – the bioremediation of a 120 km stretch of the Wadi Hanifah in Riyadh where, over the past nine years, M&T has literally brought a significant river "back from the dead." All of Moriyama & Teshima's environmental master plans have earned international recognition, significant awards and – in the case of their work on the Wadi – commendation by the United Nations.

Located in the heart of Toronto (with offices in Ottawa, Jeddah and Riyadh), M&T is an open studio that reflects the diversity of the communities they design for. Under the leadership of four principals leading the architecture and urban design group, and two principals leading the landscape architecture / environmental planning and design group, and with a total staff of 70 including architect associates and interns, landscape architects, urban planners, interior designers, graphic designers, and support staff, Moriyama & Teshima offers a full range of design and planning services, working in a deeply integrated and collaborative way to achieve unique results that capture client vision.

NKDW

attapon Klinsuwan is the principal and design director of NKDW. He was orn in 1979 in Bangkok, Thailand.

ducation: M.P.S. in Design Management, Pratt Institute, Brooklyn, New

York, 2010;

Project Featured: Chalachol Hair Salon, Bangkok 2011, SEA Thai Restaurant, Las Vegas, USA 2010

erkins + Will

erkins + Will has had a history of 77 years since it was established by wrence Perkins and Philip Will in 1935, in Chicago, United States. erkins + Will's services are very wide, including architecture, interior, ndscape, planning, construction and urban design etc. For many ears, Perkins + Will always rank the forefront of world architecture, nd no any other design companies has more projects achieving

highest LEED points in North Amercia. Perkins + Will ranks #1 in Architect Magazine in 2011.

We are honored to invite Mr. Jim Huffman, the Chief Operating Officer of Perkins + Will, to be our invited guest in this issue's interview. In the following, our topic is Green Building.

ES-Architects Ltd.

ES-Architects is one of the leading and most international chitectural design firms in Finland. Professor, Architect Pekka lminen founded the company in 1968, giving the office over 40 ars of continuous success.

e main projects of PES-Architect are such complex public buildings theatres, concert halls, airport and railway terminals, but also hools and sports facilities, retail developments and office buildings.

e partners of PES-Architects are Pekka Salminen and Tuomas vennoinen as Main Designers and Jarkko Salminen as GEO.

sides architectural design, the line of activities includes interior

design, urban planning and project management.

PES-Architects has operated in China since 2003 and in 2011 the Chinese company PES-Architects Consulting (Shanghai) Co., Ltd., P.R. China was established.

The main realized projects in Europe are Helsinki Airport in Finland and St Mary Concert Hall in Germany.

PES-Architects has participated in 60 architectural competitions in China. The Wuxi Grand Theatre was inaugurated in April 2012. The construction of the Chengdu Icon Yuan Duan super high-rise tower started at the beginning of 2011.

Pieta-Linda Auttila

Pieta-Linda Auttila, a function formed interior architect / spatial designer.

Her inspiration rests on honoring the basic essence of material, space, light and time.

She likes to survey communication between environment and human being.

By exploring fascinating features, elements and connections she wants to create systems with the most logical and flowing usabilit and thereby to enhance the natural beauty of the essentia elements of living.

In her works visual references tell their structured story abou functions, feelings, scents and sounds lead into the heart of the idea and the building starts to explain itself.

Pieta-Linda works currently as a freelance designer.

Preston Scott Cohen, Inc

Preston Scott Cohen is the Chair and Gerald M. McCue Professor of Architecture at Harvard University Graduate School of Design (GSD) and is the principal designer at Preston Scott Cohen, Inc. of Cambridge, MA. His work exemplifies a new, highly disciplined synthesis of architectural typologies, geometry and urban contexts.

Cohen's most important building to date is the Tel Aviv Museum of Art Amir Building, a 20,000 m^2 cultural facility completed in 2011. Its particular combination of galleries and exceptional public spaces is recognized for embodying the tension between two prevailing types of museums today: the museum of neutral white boxes that allows for maximum curatorial freedom and the museum of architectural exception that intensifies the experience of public spectacle.

Other important recent projects completed or under construction include the Datong City Library, the Taiyuan Museum of Art, the Nanjing Performing Arts Center, the Fahmy Residence, Los Gatos, California, and a public arcade in New York, with Pei Cobb Freed and Partners.

Awards include the Time and Leisure Best Museum of the Yea and the Design Review Award for the Tel Aviv Museum of Art five Progressive Architecture Awards, numerous first prizes in the International Competitions and an Academy Award in Architectur from the American Academy of Arts and Letters.

Cohen's work has been widely published and exhibited internationall and is the subject of numerous theoretical assessments by renowned critics and historians including Nicolai Ouroussoff, Antoine Picor Sylvia Lavin, Michael Hays, Terry Riley, Daniel Sherer, Robert Levi Robert Somol, and Rafael Moneo.

Cohen has held faculty positions at Princeton University, Rhode Islan School of Design, Ohio State University, the University of Toronto and UCLA

Renzo Piano Building Workshop

Renzo Piano was born in Genoa in 1937 into a family of builders. He developed strong attachments with this historic city and port and with his father's profession.

While studying at Politecnico of Milan University, he worked in the office of Franco Albini. After graduating in 1964, he started experimenting with light, mobile, temporary structures. Between 1965 and 1970, he went on a number of trips to discover Great Britain and the United States. In 1971, he set up the "Piano & Rogers" office in London together with Richard Rogers, with whom he won the competition for the Centre Pompidou. He subsequently moved to Paris. From the early 1970s to the 1990s, he worked with the engineer Peter Rice, sharing the Atelier Piano & Rice from 1977 to 1981. In 1981, the "Renzo Piano Building Workshop" was established, with 150 staff and offices in Paris, Genoa, and New York.

He has received numerous awards and recognitions among whic the Golden Compass Award in Milan (1981), the Royal Gold Medal the RIBA in London (1989), the Kyoto Prize in Kyoto, Japan (1990), th Neutral Prize in Pomona, California (1991), the Godwill Ambassador UNESCO (1994), the Praemium Imperiale in Tokyo, Japan (1995), th Erasmus Prize in Amsterdam (1995), the Pritzker Architecture Prize the White House in Washington (1998), the Leone d'oro alla carrie in Venice (2000), the Gold Medal of Italian architecture in Milan (200 the Gold Medal AIA in Washington 2008; and the Sonning Prize Copenhagen (2009).

Since 2004 he has also been working for the Renzo Pian Foundation, a non-profit organization dedicated to the promotic of the architectural profession through educational programs ar educational activities. The new headquarter was established in Pun Nave (Genoa), in June 2008.

Selgascano Architecture

Selgascano Architecture works in Madrid. It is a small atelier and its intention is to remain so. They have never taught at any university and they tend not to give lectures in order to avoid focusing intensely on projects. They centre their work on the construction process investigation, treating it as a continuous listening to the largest possible number of elements involved on it from manufactu to installation. They have exhibited at the MOMA in NY, th Guggenheim in NY, the Venice Biennale, the GA Gallery in Tokyo, Th MOT (Contemporany Art Museum of Tokyo) and the Design Museu of London.

Steven Holl Architects

Steven Holl Architects is a 40-person innovative architecture and urban design office working globally as one office from two locations: New York City and Beijing. Steven Holl leads the office with senior partner Chris McVoy, and junior partner Noah Yaffe. Steven Holl Architects has realized architectural works nationally and overseas, with extensive experience in campus and educational facilities, the arts (including museum, gallery, and exhibition design), and residential work. Other projects include retail design, office design, public utilities, and master planning.

Steven Holl is a tenured faculty member at Columbia University. Most recently, Holl was awarded the 2012 AIA Gold Medal, and the 2010 Jencks Award of the RIBA.

Vector Architects

Gong Dong received Bachelor & Master of Architecture from Tsinghua University, followed by a diploma at Illinois University where he received the Master of Architecture. He was also an exchange student at Technical University of Munich during his study in America. Gong Dong received several awards during his stay in America, including excellence award from Steedman Fellowship international architectural design competition in 2000, first prize from American Institute of Architects Chicago Chapter's student design competition in 2001 and excellence award from Malama Learning Centre international architecture design competition in 2002. Prior to establishing his practice he worked for Soloman Cordwell Buenz & Associates in Chicago, then at Richard Meier & Partners and Steven Holl Architects in New York. In these seven years of practices he accumulated experiences throughout conceptual design, schematic planning, construction detailing and on-site management.

During his practice as an Associate at Steven Holl Architects, Gong Dong was appointed project manager of the Linked Hybrid Building and the Vanke Center. The linked hybrid building is located in Beijing, it covers an area of 220,000 m². Environmental measures were extensively implemented in the project, such as geothermal wells, grey water recycling, radiant slab and automatic exterior shading system. The project was named one of the top ten architecture in 2006 by the American Times magazine. The Vanke centre is located in Shenzhen, it accommodates the headquarter of Vanke Company Limited. The project covers 110,000 m², and is the first LEED Platinum Rated Building in southern China.

Chien-Ho Hsu received his Master of Architecture from Harvard University. He is a member of the American Institute of Architects and a registered architect in New York State. He is also qualified as an LEED accredited professional.

During his stay in New York, Chien-Ho Hsu worked at the practices of Eisenman Architets, Gwathmey Siegel & Associates, Pei Cobb Freed & Partners and Thomas Phifer and Partners. Throughout the twelve years of practices in America he was involved in numerous projects, including IMF headquarter in Washington DC, OECD headquarter in Paris and United States Courthouse in Salt Lake City. His works covered architecture as well as urban design and interior design, driven by inclusive design and management ability and experiences. In 2010 Chien-Ho Hsu joined Vector Architects as a partner.

Vo Trong Nghia Architects

Architectural office was established by Mr. Vo Trong Nghia in the year of 2006 in Vietnam. After then, with local and foreign staffs and managers in Hanoi and Ho Chi Minh city, the office has been trying its best to devote many energy–saving architectural projects to sustainable development by using environmentally — friendly materials not only in Vietnam but also in foreign countries such as China, Cambodia, Mexico ect. Until now our result has been receiving global high appraisals and also many international awards.

4H architecture

On 1st of January 2001 Boris Zeisser and Maartje Lammers founded 4H architecture in Rotterdam, the Netherlands. Maartje Lammers was born in 1963 and graduated from Technical University of Delft in 1988. Boris Zeisser was born in 1968 and graduated from Technical University of Delft with honourable mention in 1995. Before starting their own office Boris Zeisser and Maartje Lammers worked in several well-known architectural offices like (EEA) Erick van Egeraat associated architects, Mecanoo and the Office for Metropolitan Architecture (Rem Koolhaas).

About ARTPOWER

PLANNING OF PUBLISHING

Independent plan, solicit contribution, printing, sales of books covering architecture, interior, graphic, landscape and property development.

BOOK DISTRIBUTION

Publishing and acting agency for various art design books. We support in-city call order, door to door service, mail and online order etc.

COPYRIGHT COOPERATION

To further expand international cooperation, enrich publication varieties and meet readers' multi-level needs, we stick to seeking and pioneering spirit all the way and positively seek copyright trade cooperation with excellent publishing organizations both at home and abroad.

PORTFOLIO

We can edit and publish magazine/portfolio for enterprises or design studios according to their needs.

BOOKS OF PROPERTY DEVELOPMENT AND OPERATION

We organize the publication of books about property development, providing models of property project planning and operation management for real estate developer, real estate consulting company, etc.

INTRODUCTION OF ACS MAGAZINE

ACS is a professional bimonthly magazine specializing in high-end space design. It is color printing, with 168 pages and the size of 245*325mm. There are six issues which are released in the even months every year. Featured in both Chinese and English, ACS is distributed nationwide and overseas. As the most cutting-edge counseling magazine, ACS provides readers with the latest works of the very best architects and interior designers and leads the new fashion in space design. "Present the best whole-heartedly, with books as media" is always our slogan. ACS will be dedicated to building the bridge between art and design and creating the platform for within-industry communication.

Artpower International Publishing Co., Ltd.

Add: G009, Floor 7th, Yimao Centre, Meiyuan Road, Luohu District, Shenzhen, China
Contact: Ms. Wang
Tel: +86 755 8291 3355
Web: www.artpower.com.cn
E-mail: rainly@artpower.com.cn

QR (Quick Response) Code of ACS Official Wechat Account

Acknowledgements

We would like to thank all the designers and companies who made significant contributions to the compilation of this book. Without them, this project would not have been possible. We would also like to thank many others whose names did not appear on the credits, but made specific input and support for the project from beginning to end.

Future Editions

If you would like to contribute to the next edition of Artpower, please email us your details to: artpower@artpower.com.cn